JOURNAL OF SELF-ASSEMBLY
AND MOLECULAR ELECTRONICS

Volume 1, No. 1 (January 2013)

JOURNAL OF SELF-ASSEMBLY AND MOLECULAR ELECTRONICS

Editors-in-Chief

Peter Fojan and Leonid Gurevich
Department of Physics and Nanotechnology
Aalborg University
Denmark

Aims

Self-Assembly and Molecular Electronics (SAME) is a multidisciplinary peer-reviewed journal with a wide-ranging coverage, specializing in the areas of molecular electronics and self assembly systems. SAME encourages original cross-disciplinary full research articles, rapid communications of important scientific and technological findings, and state-of-the-art reviews.

Scope

SAME publishes theoretical and experimental original research covering areas of:

- Molecular Electronics and Molecular Devices with a particular emphasis on DNA, peptide and protein based systems
- Self Assembly in Nanosience, Chemistry, Biology and Medicine
- Supramolecular Chemistry
- Modelling of Structural and Electronical Properties of Organic Molecules and Self Assembled Systems

JOURNAL OF SELF-ASSEMBLY
AND MOLECULAR ELECTRONICS

Volume 1 No. 1 January 2013

Published, sold and distributed by:
River Publishers
P.O. Box 1657
Algade 42
9000 Aalborg
Denmark

Tel.: +45369953197
www.riverpublishers.com

Journal of Self-Assemly and Molecular is published three times a year.
Publication programme, 2013: Volume 1 (3 issues)

ISSN 2245-4551
ISBN 978-87-92982-38-4 (this issue)

Editorial Foreword:
Welcome to the Journal of Self-Assembly and Molecular Electronics (SAME)

Dear Readers,

Welcome to our new cross-disciplinary journal devoted to self-assembly and molecular electronics. These are the two pillars that are required to fulfill the dream of bottom-up nanotechnology. This journal will be open to both theoretical and experimental works coming from all walks of science and engineering: Physics, Chemistry, Biology, Material Science, Mechanical, Electrical and Medical Engineering, Bio- and Nanotechnology. We envisage the journal as a part of a much broader platform with the aim to create a stimulating environment for interdisciplinary communications and thriving scientific community in this exciting field of science. Encompassing the journal contributions, the biennial conference on this subject, regular schools and workshops, this platform will provide an inspiring and broad discussion forum to promote and advance this field of science.

The first issue, which we are proud to present, addresses the topic of DNA-based molecular electronics. DNA is not only the corner stone of life on earth carrying the assembly instructions for all living organisms, but also proved to be an amazing tool in nanotechnology. DNA-arrays and DNA-origami demonstrate how the assembly of this molecule into a particular shape can be encoded into its sequence. On the other hand, the potential application of DNA building blocks for molecular electronics highly depends on its long-range charge transfer properties which are still highly debated. This journal issue presents several research papers and reviews on theoretical and experimental methods required to construct such DNA-based devices. We are indebted to our authors who trusted our vision and submitted their work to be published in the first issue of the journal.

We invite you, our readers, to contribute to this vision with your scientific work. We, on our side, guarantee fast, fair and unbiased peer-review and high quality publication service both online and in print.

Leonid Gurevich and Peter Fojan
Editors

Modeling Charge Transport and Dynamics in Biomolecular Systems

R. Gutierrez[1] and G. Cuniberti[1,2]

[1]Institute for Materials Science and Max Bergmann Center of Biomaterials,
Dresden University of Technology, 01062 Dresden, Germany
[2]Division of IT Convergence Engineering and National Center for Nanomaterials
Technology, POSTECH, Pohang 790-784, Republic of Korea;
e-mail: projects@nano.tu-dresden.de

Received 3 December 2012; Accepted 10 December 2012

Abstract

Charge transport at the molecular scale builds the cornerstone of molecular electronics (ME), a novel paradigm aiming at the realization of nanoscale electronics via tailored molecular functionalities. Biomolecular electronics, lying at the borderline between physics, chemistry and biology, can be considered as a sub-field of ME. In particular, the potential applications of DNA oligomers either as template or as active device element in ME have strongly drawn the attention of both experimentalist and theoreticians in the past years. While exploiting the self-assembling and self-recognition properties of DNA based molecular systems is meanwhile a well-established field, the potential of such biomolecules as active devices is much less clear mainly due to the poorly understood charge conduction mechanisms. One key component in any theoretical description of charge migration in biomolecular systems, and hence in DNA oligomers, is the inclusion of conformational fluctuations and their coupling to the transport process. The treatment of such a problem affords to consider dynamical effects in a non-perturbative way in contrast to, e.g., conventional bulk materials. Here we present an overview of recent work aiming at combining molecular dynamic simulations and electronic structure calculations with charge transport in coarse-grained effective model

Journal of Self-Assembly and Molecular Electronics, Vol. 1, 1–39.

Hamiltonians. This hybrid methodology provides a common theoretical starting point to treat charge transfer/transport in strongly structurally fluctuating molecular-scale physical systems.

Keywords: Electronic structure, biomolecules, molecular dynamics, quantum transport.

1 Introduction

Can a DNA molecular wire mediate charge transport, i.e., support an electrical current? This question has attracted a considerable attention in the past 20 years, mainly triggered by two facts: first, the emergence of molecular electronics [1–4], which opened the perspective of realizing electronic functions at the molecular level and thus of exploiting DNA self-assembling and self-recognition properties [5–7]; second, by the demonstration of long-range hole transfer in different DNA oligomers in solution in the groups of Barton [8–13], Giese [14, 15], Lewis [16], Schuster [17–19], and Michel-Beyerle [14, 20]. These experiments revealed electron transfer occurring over distances as long as 200 Å, which was in so far surprising, as the highly disordered structure of natural DNA – related to a quasi-random base pair sequence – should lead to a considerable degree of charge localization. Mechanisms based on thermally activated incoherent hole hopping have been successful in describing hole transfer in solution [20, 28–30].

However, when coming to investigate experimentally charge transport properties, i.e. the electrical response of DNA when contacted by metallic electrodes, the situation became considerably less clear. Indeed, a variety of partially contradictory experimental results has been obtained in the past years for double-strand (ds) DNA oligomers [19, 21–27]. This situation hints not only at the difficulties encountered to carry out well-controlled transport measurements, but also at the strong sensitivity of charge migration to several factors, including (i) the specific base sequence of the probed molecules, (ii) the structure of DNA (whether in A or B form) which can lead to a quantitative difference in the overlap of the π-electrons of the stacked bases, (iii) the length of the sequences – longer chains may be deformed due to structural instability – so that any kinks and defects in the DNA structure introduced in this way may distort the DNA π-system, and (iv) the DNA-metal contact topology and electronic structure.

Meanwhile, some experiments [25, 26, 39] have shown in a reliable way that charge transport can take place through single short DNA segments. Xu et al. [25] and Cohen et al. [26] addressed dsDNA while Liu et al. [39] focused on G4. Concerning the dsDNA, both experiments found currents of 100–150 nA at about 1 V. In [25], electrical transport through covalently Au-contacted double-stranded DNA in aqueous solution was measured, where the native form of the DNA is preserved. The $I–V$ curves show a rather smooth ohmic profile with considerably large currents up to 150 nA at 0.8 V. The experimental approach of Cohen et al. [26] was based on measuring current through suspended dsDNA molecules connected between a metal substrate and a gold nanoparticle contacted to an AFM conducting tip. Currents of the order of 220 nA at 2.0 V were measured. Interestingly, the length and base sequence of DNA in these two experiments were completely different. Unfortunately, *systematic* investigations (within a given experimental setup) on base sequence, length, and the temperature dependence of charge transport are still missing, so that the theoretical analysis of possible charge transport pathways faces big challenges [40]. Initial modeling of charge transport was mainly based on effective Hamiltonians with fixed electronic parameters and describing e.g hole transport through the highest occupied molecular orbitals (HOMO) of the bases [41–47] (see also [48,49] for recent reviews). Model Hamiltonians clearly offer the possibility to explore different charge transport scenarios in a relatively flexible way, but they also contain usually many parameters which are difficult to estimate. This limits the predictive power of those models in case of complex biomolecular systems. On the other hand, first-principle calculations [50–63] performed on static structures can provide accurate values for the electronic couplings but can hardly deal with the full transport problem. As a result, the development of methodologies exploiting the advantages of both approaches become highly desirable.

An important factor governing charge transfer/transport is the biomolecular structural dynamics: several important studies in the chemical physics community have highlighted the crucial role played by dynamical fluctuations in favoring or hindering hole migration [30–38, 61, 64–72]. Hence, we may expect that transport of charges when DNA is contacted by electrodes can only be understood in the context of a *dynamical approach* which includes the coupling of the electronic system to conformational degrees of freedom, resulting in fully or partially incoherent charge propagation. Indeed, strong conformational dynamics and a related spectrum of different time scales seem to be ubiquitous for biomolecules, as shown e.g., in the modulation of the kinetics of electron transfer during the early stages of the photosynthetic reaction cycle [73], by the analysis of dispersed kinetics [74,75], or by fluorescense correlation spectroscopy [76]. The treatment of dynamical effects in DNA transport calculations has been, however, addressed only recently either in the frame of pure model Hamiltonians [47, 77–80] or by including information from first principle calculations and molecular dynamics (MD) simulations [67, 69, 81–84].

A realistic inclusion of the influence of dynamical effects onto the transport properties can however only be achieved in our view via hybrid methodologies combining a reliable description of the biomolecular dynamics and electronic structure with quantum transport calculations. The present paper will introduce such a methodology which has been recently developed [81–83, 85–88]. This approach allows to consider the influence of structural fluctuations and solvent effects onto the electronic structure of DNA oligomers. Hereby we use a density-functional (DFT) based fragment-orbital method, which provides a very efficient way to compute the charge transfer parameters along nanosecond molecular dynamic trajectories. Solvation effects are described using a hybrid quantum mechanics/molecular mechanics (QM/MM) coupling scheme. The combination of the method with coarse-grained Hamiltonian models opens the way to study charge transport in complex systems where the interaction with dynamical degrees of freedom plays a fundamental role.

The paper is organized as follows. In the next section we present a model Hamiltonian approach including coupling to specific vibrational degrees of freedom that was used to describe charge transport through standing DNA sequences [26]. The goal of the section is twofold: first to introduce Green function techniques and to show how to treat analytically such problems; second to illustrate what are the intrinsic limitations of approaches based only on model Hamiltonians. In Section 3 we shall then to introduce the computational methodology combining classical molecular dynamic simulations with DFT based electronic structure calculations along the MD trajectory. An efficient coarse-graining is hereby achieved by using fragment orbitals. The resulting effective electronic structure can be mapped onto a linear tight-binding chain with time-dependent electronic parameters. Charge transport can be then investigated along two different but complementary ways: by computing quantum mechanical transmission probabilities along the MD trajectory and performing time averages (Section 3.2) or by mapping the time dependent electronic Hamiltonian onto a model describing the interaction of a static (time averaged) electronic system with a dissipative bosonic environment (Section 4). The main advantage of this second approach is the possibility to describe charge transport beyond the coherent limit. More importantly, the quantities characterizing the bath (spectral density) can be fully determined in terms of time correlation functions of the electronic parameters.

2 Model Hamiltonian Approaches to Charge Transport in DNA Molecular Wires: Standing DNA Oligomers

In the context of charge transport, model Hamiltonian formulations have been extensively applied to address the role played by different physical parameters in determining the efficiency of charge migration through different DNA oligomers. Cuniberti et al. [48] offer a recent overview of different classes of such tight-binding based models commonly used in the past years to compute the charge transport characteristics of DNA wires.

In this section we will present an example on how a model Hamiltonian approach can be used to describe some experimental findings. We pursue hereby two main goals: (i) to show how analytical methods based on Green functions can be applied to deal with the coupling to vibrational excitations in a typical minimal model, and (ii) to demonstrate what is the main drawback of a model treatment of transport, namely, the appearance of a number of free parameters, which are in general difficult to obtain from first-principle calculations. This second aspect is crucial since the complexity of DNA molecules makes a full ab initio estimation of such parameters very challenging and the possible transport scenarios do dramatically depend on the specific values these parameters can take.

The reference point for the theoretical treatment were single molecule experiments performed at the group of D. Porath [26]. In brief, a thiolated 26 bases long single-stranded DNA (ssDNA) with a non-homogeneous sequence was adsorbed on a gold surface to create a dense monolayer. The ssDNA molecules had a thiol-modified linker end group $(CH_2)_3$-SH at the 3'-end (see Figure 1(a)). The monolayer was then exposed to a solution of the complimentary ssDNA bound to gold nanoparticles (GNP) via a thiol linker. Upon incubation of ssDNA-GNP conjugates on ssDNA- functionalized gold surface, a certain density of dsDNA attached at one end to the gold surface and at the other end to GNP was formed. A metal coated AFM tip was used to form an electrical contact to protruding GNP and perform electrical measurements across dsDNA. Figure 1(b) shows an AFM image of several GNPs, indicating the position of the hybridized dsDNA on the background of the ssDNA monolayer. The recorded current voltage curves, shown in Figure 1(c)

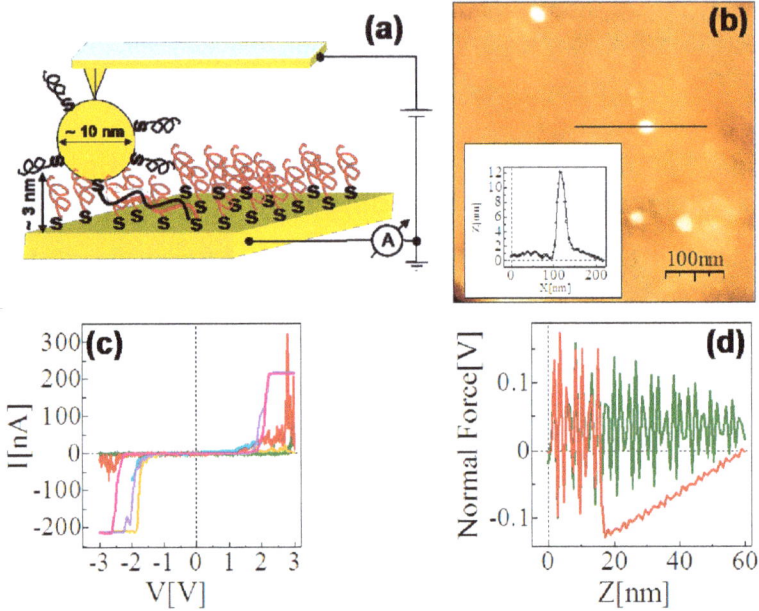

Figure 1 (a) Scheme of the experimental setup showing dithiolated dsDNA chemically bonded to two metal electrodes (upper – GNP, lower – gold surface). The base sequence is given by 5'-CAT TAA TGC TAT GCA GAA AAT CTT AG-3'-(CH$_2$)$_3$-SH. (b) AFM topography image showing a top view of the sample. The GNPs mark the position of the hybridized dsDNA. (c) Collection of I–V curves from different samples. (d) An F–Z curve of one of the curves in (c), (green – forward, red – backward) demonstrating the tip-GNP adhesion (red line) without pressing the GNP through the monolayer. Reprinted with permission from [79]. © 2006 American Physical Society (DOI: 10.1103/PhysRevB.74.235105).

demonstrate in a clear and reproducible way, the ability of ∼9 nm long dsDNA to conduct relatively high currents (> 200 nA), when the molecule is not attached to a hard surface along its backbone and when charge can be injected efficiently through a chemical bond. Such behavior was measured for many dsDNA molecules on tens of samples and with various tips and humidity conditions with similar results.

From a theoretical point of view, the central aim is to formulate a *minimal* model describing the DNA electronic structure and the coupling to the electrodes (metallic surface and GNP modified AFM tip) [79]. For this, we adopt the perspective that to describe low-energy quantum transport within a single-particle picture, only the frontier π orbitals of the base pairs are relevant. The starting point will be then a planar ladder model with a single orbital

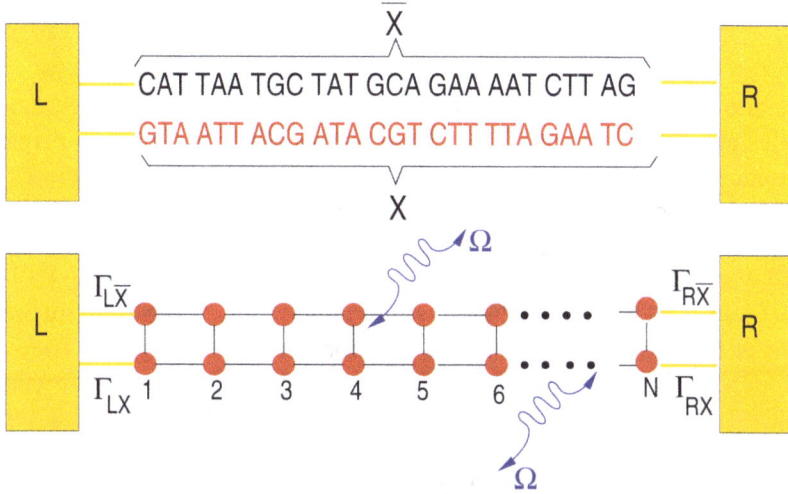

Figure 2 Upper panel: Schematic representation of the double-strand DNA with the experimentally relevant base-pair sequence [26]. The $(CH_2)_3$-SH linker groups are omitted for simplicity (see the text for details). Lower panel: Two-strand ladder used to mimic the double-strand structure of a DNA molecule. L and R refer to left and right electrodes, respectively. The coupling terms to the electrodes $\Gamma_{\ell,\alpha}$, $\ell = X, \bar{X}$, $\alpha = L, R$ are assumed to be energy-independent constants. Reprinted with permission from [79]. © 2006 American Physical Society (DOI: 10.1103/PhysRevB.74.235105).

per lattice site within a nearest-neighbor tight-binding picture as shown Figure 2. Ladder models have been previously used by different authors to study quantum transport in DNA duplexes [43, 46, 89–92]. The Hamilton operator describing the ladder and its coupling to left (L) and right (R) electronic reservoirs is given by

$$
\mathcal{H}_{el} = \sum_{r=X,\bar{X}} \sum_{\ell} \epsilon_{r,\ell} b_{r,\ell}^{\dagger} b_{r,\ell}
$$

$$
- \sum_{r=X,\bar{X}} \sum_{\ell} t_{r,\ell,\ell+1} [b_{r,\ell}^{\dagger} b_{r,\ell+1} + \text{h.c.}]
$$

$$
- \sum_{\ell} t_{\perp,\ell} [b_{X,\ell}^{\dagger} b_{\bar{X},\ell} + \text{h.c.}]
$$

$$
+ \sum_{k \in L} [t_{k,X} c_{k}^{\dagger} b_{X,1} + \text{h.c.}] + \sum_{k \in L} [t_{k,\bar{X}} c_{k}^{\dagger} b_{\bar{X},1} + \text{h.c.}]
$$

$$+ \sum_{k \in R} [t_{\mathbf{k},X} c_{\mathbf{k}}^{\dagger} b_{X,N} + \text{h.c.}] + \sum_{k \in R} [t_{\mathbf{k},\bar{X}} c_{\mathbf{k}}^{\dagger} b_{\bar{X},N} + \text{h.c.}] \qquad (1)$$

In the above expression, $X, \bar{X}$ refer to the two strands of the ladder, $\epsilon_{r,\ell}$ are energies at site ℓ on strand r, $t_{r,\ell,\ell+1}$ are the corresponding nearest-neighbor electronic hopping integrals along the two strands while $t_{\perp,\ell}$ describes the inter-strand hopping. To have an estimation of the onsite energies, values obtained by Mehrez and Anantram [60] using density-functional theory (DFT) were taken. Considering electron transport, we chose $\epsilon_G = 1.14$ eV, $\epsilon_C = -1.06$ eV, $\epsilon_A = 0.26$ eV, $\epsilon_T = -0.93$ eV. More difficult is the choice of the intra- and inter-strand electronic transfer integrals. They will be more sensitive to the specific base sequence considered. For the sake of simplicity and in order to reduce the number of model parameters a simple parameterization is adopted taking homogeneous hopping along both strands, i.e. $t_{r,\ell,\ell+1} = t_X = t_{\bar{X}} = t \sim 0.25$–$0.27$ eV and $t_{\perp,\ell} = t_{X\bar{X}} \sim 0.2$–$0.3$ eV. We remark that the hopping integrals are considered as effective parameters, thus keeping some freedom in the choice of their specific values. Electronic correlations [46] or structural fluctuations mediated by the coupling to other vibrational degrees of freedom (see the next paragraphs) can lead to a strong renormalization of the bare electronic coupling. The interaction with the electronic reservoirs will be described in the most simple way by invoking the wide-band approximation, i.e. neglecting the energy dependence of the lead self-energies (see below).

To model the coupling to vibrational degrees of freedom we consider the case of long-wave length optical modes with constant frequencies Ω_α, e.g., long wave-length torsional modes and assume they couple to the total charge density operator $N = \sum_{r,\ell} n_{r,\ell} = \sum_{r,\ell} b_{r,\ell}^{\dagger} b_{r,\ell}$ of the ladder. This approximation can be justified for long-wave length distortions. In other words, the strength of the electron-vibron interaction λ is assumed to be site-independent. The total Hamiltonian thus reads [79]:

$$\mathcal{H} = \mathcal{H}_{\text{el}} + \sum_{\alpha} \Omega_\alpha B_\alpha^{\dagger} B_\alpha + \sum_{r,\ell,\alpha} \lambda_\alpha b_{r,\ell}^{\dagger} b_{r,\ell} (B_\alpha + B_\alpha^{\dagger}) \qquad (2)$$

To deal with the transport problem using this Hamiltonian model we will use Green function techniques which offer a powerful methodology to cover different transport regimes as well as to treat the coupling between different degrees of freedom. In a first step, a Lang–Firsov (LF) unitary transformation [93] is performed in order to remove the electron-vibron interaction. The LF-generator is given by $\mathcal{U} = \exp[-\sum_{\alpha,r,\ell} g_\alpha b_{r,\ell}^{\dagger} b_{r,\ell} (B_\alpha - B_\alpha^{\dagger})]$, which is basically a shift operator for the harmonic oscillator position. The parameter $g_\alpha = \lambda_\alpha/\Omega_\alpha$ gives an effective measure of the electron-vibron coupling strength. As a result of the LF transformation, the onsite energies $\epsilon_{r,\ell}$ are shifted to $\epsilon_{r,\ell} - \Delta$ with $\Delta = \sum_\alpha \lambda_\alpha^2/\Omega_\alpha$ being the so called polaron shift. Renormalization effects in the tunneling Hamiltonian will be neglected in the following.

For the transport problem, the standard current expression for lead $p=L,R$ as derived by Meir and Wingreen [94] can be used:

$$I_p = \frac{2i\,e}{h} \int dE \, \mathrm{Tr}[\Gamma_p\{f_p(G^> - G^<) + G^<\}], \tag{3}$$

and then the LF unitary transformation is performed under the trace going over to transformed Green functions. In the previous equation, $\Gamma_p(E) = i\,(\Sigma_p(E) - \Sigma_p^\dagger(E))$ are the lead spectral functions, $f_p(E) = f(E - \mu_p)$ is the Fermi function of the p-lead and $\mu_{p=L} = E_F + eV/2$ ($\mu_{p=R} = E_F - eV/2$) are the corresponding electrochemical potentials. Within the wide-band limit in the electrode coupling, the following $2N \times 2N$ ladder-lead energy-independent coupling matrices can be defined:

$$(\Gamma_L)_{nm} = \begin{cases} \Gamma_{L,X}\delta_{n,1}\delta_{m,1} & \text{if } n,m \in X \\ \Gamma_{L,\bar{X}}\delta_{n,1}\delta_{m,1} & \text{if } n,m \in \bar{X} \\ 0 & \text{if } \text{elsewhere} \end{cases}$$

$$(\Gamma_R)_{nm} = \begin{cases} \Gamma_{R,X}\delta_{n,N}\delta_{m,N} & \text{if } n,m \in X \\ \Gamma_{R,\bar{X}}\delta_{n,N}\delta_{m,N} & \text{if } n,m \in \bar{X} \\ 0 & \text{if } \text{elsewhere} \end{cases}$$

Now define the fermionic vector operator (see Figure 2 for reference):

$$\Psi^\dagger = (b_{X,1}\,b_{X,2}\cdots b_{X,N}\,b_{\bar{X},1}\cdots b_{\bar{X},N}). \tag{4}$$

Lesser- and greater-matrix Green functions (GF) can then be introduced as

$$G^>(t) = -\frac{i}{\hbar}\langle\Psi(t)\Psi^\dagger(0)\rangle,$$

$$G^<(t) = \frac{i}{\hbar}\langle\Psi^\dagger(0)\Psi(t)\rangle. \tag{5}$$

Since Eq. (3) does not explicitly contain information on the specific structure of the "molecular" Hamiltonian, we can now transform the lesser- and

greater-GF as well as the lead spectral functions to the polaron repres-
entation. The operator Ψ transforms according to $\bar{\Psi} = \mathcal{U}\Psi\mathcal{U}^\dagger = \Psi\mathcal{X}$,
where $\mathcal{X} = \exp[\sum_\alpha (\lambda_\alpha/\Omega_\alpha)(B_\alpha - B_\alpha^\dagger)]$. Thus, we obtain $\bar{G}^>(t) =$
$-(i/\hbar)\langle\Psi(t)\mathcal{X}(t)\,\Psi^\dagger(0)\mathcal{X}^\dagger(0)\rangle$ and similar for $\bar{G}^<(t)$. Performing an approximate decoupling in this expression into polaronic and vibronic com-
ponents one obtains:

$$\bar{G}^>(t) = -\frac{i}{\hbar}\langle\Psi(t)\mathcal{X}(t)\,\Psi^\dagger(0)\mathcal{X}^\dagger(0)\rangle$$

$$\approx -\frac{i}{\hbar}\langle\Psi(t)\Psi^\dagger(0)\rangle_{el}\langle\mathcal{X}(t)\mathcal{X}^\dagger(0)\rangle_B$$

$$= G^>(t)\langle\mathcal{X}(t)\mathcal{X}^\dagger(0)\rangle_B = G^>(t)e^{-\Phi(t)}, \tag{6}$$

with a similar expression holding for the lesser-than GF by changing the time
argument t by $-t$ in $\Phi(t)$.

For a single vibrational mode, the vibron correlation function $\Phi(t)$ can be
exactly evaluated [93]:

$$e^{-\Phi(t)} = e^{-g^2(2N+1)}\sum_{n=-\infty}^{\infty} I_n(\tau)e^{\beta\Omega n/2}e^{-i\,n\Omega t}, \tag{7}$$

where $\tau = 2g^2\sqrt{N(N+1)}$ and $g = \lambda/\Omega$. It follows then for the Fourier
transformed lesser and greater GFs:

$$\bar{G}^{<(>)}(E) = \sum_{n=-\infty}^{\infty}\phi_n(\tau)G^{<(>)}(E + (-)n\Omega),$$

$$\phi_n(\tau) = e^{-g^2(2N+1)} \times I_n(\tau)\,e^{\beta\Omega n/2}. \tag{8}$$

where $+(-)$ corresponds to $< (>)$. The bare lesser- and greater-GF can now
be obtained from the kinetic equation $G^{<(>)} = G^r(\Sigma_L^{<(>)} + \Sigma_R^{<(>)})G^a$, since
the full electron-vibron coupling is already contained in the prefactor function
$\phi_n(\tau)$. The leads self-energy matrices $\Sigma_p^<$, $\Sigma_p^>$ are given in the wide-band
limit by $i\,f_p(E)\Gamma_p$ and $-i\,(1 - f_p(E))\Gamma_p$, respectively. Using these expressions, the total symmetrized current in the stationary state $I_T = (I_L - I_R)/2$
is given by [79]

$$I_T = \frac{e}{2h}\sum_{n=-\infty}^{\infty}\phi_n(\tau)\int dE\,\{[f_L(E)\,(1 - f_R(E - n\Omega))$$

$$- f_R(E)(1 - f_L(E - n\Omega))] t(E - n\Omega)$$
$$+ [f_L(E + n\Omega)(1 - f_R(E))$$
$$- f_R(E + n\Omega)(1 - f_L(E))] T(E + n\Omega)\}, \tag{9}$$

where $T(z) = \text{Tr}[\Gamma_R G^r(z) \Gamma_L G^a(z)]$ is the conventional expression for the transmission coefficient in terms of the molecular Green function $G(E)$, which satisfies the Dyson-equation: $G^{-1} = G_0^{-1} - \Sigma_L - \Sigma_R$. The above result for the current has a clear physical interpretation. So, e.g., a term like $f_L(E)(1 - f_R(E - n\Omega))T(E - n\Omega)$ describes an electron in the left lead which tunnels into the molecular region, emits n vibrons of frequency Ω and tunnels out into the right lead. However, it can only go into empty states, hence the Pauli blocking factor $(1 - f_R(E - n\Omega))$. Other terms can be interpreted along the same lines, when one *additionally* substitutes electrons by holes.

Finally, a spectral density $A(E, V)$ can be defined as

$$A(E, V) = i\,[\bar{G}^>(E) - \bar{G}^<(E)]$$
$$= i \sum_n \phi_n(\tau)\,[G^>(E - n\Omega) - G^<(E + n\Omega)]. \tag{10}$$

Figure 3(a) shows the electronic band structure of an *infinite* periodic array of the 26-base-pairs DNA molecule without considering charge-vibron interactions. The strongly fragmented energy spectrum is a result of the small hopping integrals and the inhomogeneous onsite energy distribution. We may thus rather speak of valence and conduction manifolds as of true dense electronic bands [56]. In Figure 3(a) we also show schematically the positions of the conduction and valence manifolds of a periodic poly(GC) reference system (open rectangles). Figures 3(b)–(d) show the spectral density at zero voltage of the *finite* DNA ladder contacted by electrodes in three different ways: (b) only the 3'-ends, (c) only the 5'-ends, and (d) all four ends of the double-strand are contacted. Though the general effect consists in broadening of the electronic manifolds, we also see that depending on the way the molecule is contacted to the leads the electronic states will be affected in different ways. Thus, e.g., states around 1.7 eV above the Fermi level are considerably more broadened than states closer to E_F.

Considering now the coupling to vibrational degrees of freedom in the ladder, one should notice that the probability of opening inelastic transport channels by emission or absorption of n vibrons becomes higher with increasing thermal energy $k_B T$ and/or electron-vibron coupling g. As a result, the

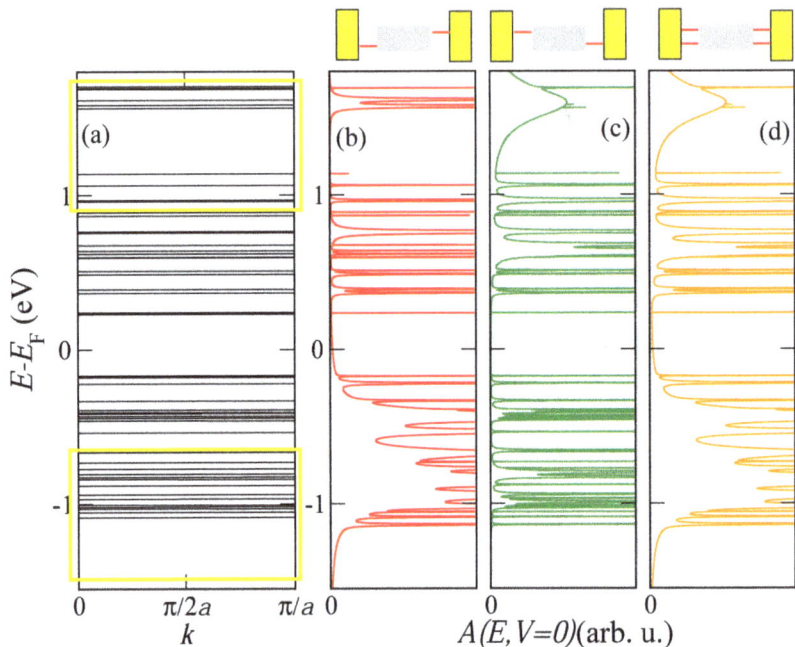

Figure 3 (a) Tight-binding electronic band structure of an infinite DNA system, obtained by a periodic repetition of the 26-base sequence of Cohen et al. [26]. The open yellow rectangles indicate for reference the approximate position of the bands for a periodic poly(GC) oligomer. (b)–(d) Spectral density $A(E, V = 0)$, which at zero voltage coincides with the projected density of states onto the molecular region, for the *finite size* DNA chain contacted in different ways by left and right electrodes (see Figure 2). Reprinted with permission from [79]. © 2006 American Physical Society (DOI: 10.1103/PhysRevB.74.235105).

spectral density $A(E)$ will consist of a series of elastic peaks (corresponding to $n = 0$) plus vibron satellites ($n \neq 0$).

Figure 4 shows the influence of the coupling to the vibron mode on the magnitude of the current and of the zero-current gap. The slope of the I–V curves is considerably reduced with increasing g. The corresponding spectral densities at $V \sim 1.5$ V (see Figure 4, lower panel) show broadening due to the emergence of an increasing number of vibron satellites (inelastic channels) with larger coupling, but at the same time a redistribution of spectral weights takes place. This is simply the result of a sum rule $\int dE \, A(E) = 2\pi$. The reason for the current reduction can be qualitatively understood by looking at the spectral density. The reduction in the intensity of $A(E)$ will clearly lead to a reduction in the current at a fixed voltage, since it is basically the area

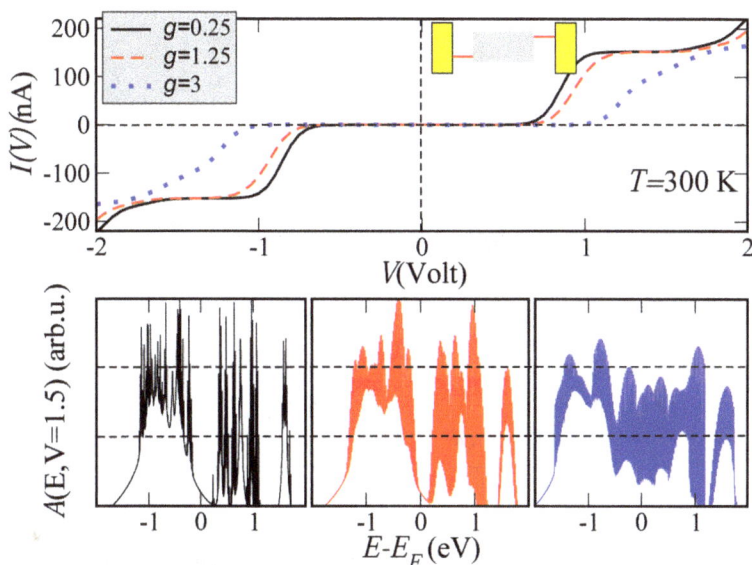

Figure 4 Dependence of the current on the effective electron-vibron coupling strength $g = \lambda/\Omega$ at room temperature. With increasing coupling the total current is reduced and the zero-current gap is enhanced. The lower panels show the spectral density at a fixed bias voltage $V \sim 1.5$ V for different values of g. Reprinted with permission from [79]. © 2006 American Physical Society (DOI: 10.1103/PhysRevB.74.235105).

under $A(E, V = \text{const.})$ within the energy window $[E_F - eV/2, E_F + eV/2]$ which really matters. Notice also the increase of the zero-current gap with increasing electron-vibron coupling (vibron blockade), which is related to the exponential suppression of transitions between low-energy vibronic states [95]. Alternatively, this can be interpreted as an increase of the effective mass of the polaron which thus leads to its localization and to a blocking of transport at low energies.

The measured I–V characteristics [26] can be described semi-quantitatively by reformulating the previous model to include two vibrational excitations. The extension of the model is straightforward and details can be found in [79]. Figure 5 shows two experimental curves and the corresponding theoretical I–V plots. The values used for the charge-vibron coupling ($\lambda_1 = 15(35)$ meV, $\lambda_2 = 15(20)$ meV) and vibron frequencies ($\Omega_1 = 20$ meV, $\Omega_2 = 6$ meV) for the yellow (black) theoretical curves have reasonable orders of magnitude for low-frequency modes, see e.g. [96].

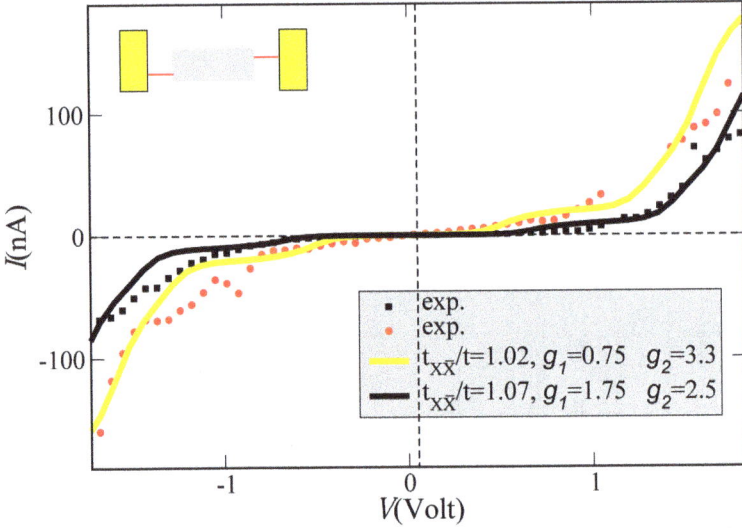

Figure 5 Theoretical curves (solid lines) compared with two different I–V curves as obtained on suspended double-strand DNA oligomers contacted by a GNP [26]. In both cases the temperature and the coupling to the electrodes were kept fixed at $T = 300$ K and $\Gamma_{L,X} = \Gamma_{R,\bar{X}} = 250$ meV, $\Gamma_{R,X} = \Gamma_{L,\bar{X}} = 0$, respectively. Reprinted with permission from [79]. © 2006 American Physical Society (DOI: 10.1103/PhysRevB.74.235105).

The theoretical approach presented in this section highlights the advantages of a model based formulation: (i) low computational cost, (ii) the possibility of obtaining analytical results to analyze limiting cases, and (iii) to cover different physical regimes, like e.g., strong or weak coupling to vibrational modes. However, also the main drawback of a pure model Hamiltonian approach to transport in complex systems can be clearly appreciated: the results can strongly depend on the parameter choice, the number of parameters usually growing with the increase in complexity of the models. Since a purely ab initio based description of transport in complex biomolecular systems is not feasible, it is thus desirable to develop methodologies able to bridge the gap between reliable electronic structure calculations and transport models. The next section will be devoted to present such an approach.

3 Structural Fluctuations in Biomolecular Systems: Bridging Molecular Dynamics with Transport Models

In this section, a methodology [81–83, 85–88] is described which uses a hybrid approach based on a combination of molecular dynamics simulations and electronic structure calculations with a mapping of the time-fluctuating electronic structure along the MD trajectory onto coarse-grained transport models. The key issue is that in this way the number of free parameters in the model formulation is reduced to a large extent. Moreover, the degree of coarse-graining by the formulation of the charge transport model can be progressively improved in a controlled way by adding stepwise information drawn from the electronic structure calculations. In the following the transport problem will be approached from two complementary points of view. In the first one charge transport will be addressed via the computation of time averaged transmission functions, i.e., at regular snapshots t_i along the MD trajectory the electronic structure is calculated and from there a quantum mechanical transmission $T(E, t_i)$ is computed, thus generating a time dependent $T(E, t_i)$ which is then averaged over the simulation time. In the second approach, a model is formulated which describes the coupling of the electronic system to a bosonic bath which comprises internal vibrations and solvent effects. The bath encodes the dynamical information drawn from the MD simulations. The bath spectral density can then be calculated from time series generated during the MD run.

3.1 Electronic Structure and Fragment Orbital Approach

We will first describe the strategy used to calculate in a very efficient way the electronic structure of an arbitrary DNA sequence (the method can be of course applied to other biomolecular systems or to organic stacks) [81–83, 85–88]. The immediate goal is to map the electronic structure onto a tight-binding model with time-dependent electronic coupling and onsite energies. The tight-binding Hamiltonian takes the form

$$H = \sum_i \epsilon_i a_i^\dagger a_i + \sum_{ij} V_{ij}(a_i^\dagger a_j + \text{h.c.}). \qquad (11)$$

The onsite energies ϵ_i and the nearest-neighbor hopping integrals V_{ij} characterize, respectively, effective ionization energies and electronic couplings of the molecular fragments (see below). The evaluation of these parameters is done by using the SCC-DFTB method [97] combined with a fragment orbital

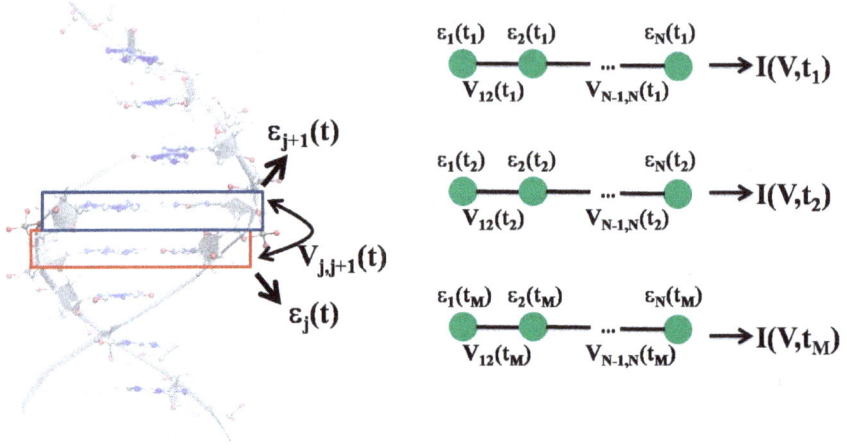

Figure 6 Left panel: Schematic representation of the fragment orbital method used to perform a coarse-graining of the DNA electronic structure. A fragment consists of a single base pair (not including the sugar phosphate backbones). As explained in the text, the hopping matrix elements $V_{j,j+1}$ between nearest-neighbor fragments are computed using the molecular orbital basis of the isolated base pairs. These calculations are then carried out at snapshots along the molecular dynamics trajectory hence leading to time dependent electronic structure parameters. By keeping only one relevant orbital per fragment, the electronic structure can be mapped onto that of a linear chain (right panel). Reprinted with permission from [83]. © 2010, Institute of Physics.

(FO) approach [87, 88, 98]:

$$\epsilon_i = -\langle \phi_i | \hat{H}_{KS} | \phi_i \rangle \tag{12}$$

and

$$V_{ij} = \langle \phi_i | \hat{H}_{KS} | \phi_j \rangle. \tag{13}$$

The molecular orbitals ϕ_i and ϕ_j are, e.g., the highest-occupied molecular orbitals (HOMO) of the DNA bases i and j. Depending on the definition of the FOs, different tight-binding models may be designed. In our case, we use a minimal approach where the DNA electronic structure will be mapped onto a linear chain. The FOs are obtained by performing SCC-DFTB calculations of the isolated fragments, i.e., the individual nucleotide pairs in this case.

Performing an LCAO expansion

$$\phi_i = \sum_\mu c_\mu^i \eta_\mu, \tag{14}$$

the coupling and overlap integrals in the molecular-orbital (MO) basis can be evaluated as

$$V_{ij} = \sum_{\mu\nu} c_\mu^i c_\nu^j \langle \eta_\mu | \hat{H}_{KS} | \eta_\nu \rangle = \sum_{\mu\nu} c_\mu^i c_\nu^j H_{\mu\nu} \qquad (15)$$

and

$$S_{ij} = \sum_{\mu\nu} c_\mu^i c_\nu^j \langle \eta_\mu \mid \eta_\nu \rangle = \sum_{\mu\nu} c_\mu^i c_\nu^j S_{\mu\nu}. \qquad (16)$$

$H_{\mu\nu}$ and $S_{\mu\nu}$ are the Hamilton and overlap matrices in the atomic basis set as evaluated with the SCC-DFTB method [87].

The effect of environment, i.e. the electrostatic field of the DNA backbone, the water molecules and the counter-ions, is taken into account through the following QM/MM Hamiltonian:

$$H_{\mu\nu} = H_{\mu\nu}^0 + \frac{1}{2} S_{\mu\nu}^{\alpha\beta} \sum_\delta \Delta q_\delta (\gamma_{\alpha\delta} + \gamma_{\beta\delta}) + \sum_A Q_A \left(\frac{1}{r_{A\alpha}} + \frac{1}{r_{A\beta}} \right) \qquad (17)$$

Δq_δ are Mulliken charges in the QM region and Q_A are charges in the MM region, i.e. the DNA backbone, counter-ions and water molecules. The coupling to the environment is therefore explicitly described via the interactions with the Q_A charges. In the following, the calculation scheme based on the complete expression in Eq. (17) will be denoted as QM/MM, while neglecting the last term will be denoted as "vacuo". The matrix V_{ij} is built from non-orthogonal orbitals ϕ_i and ϕ_j, so that it will renormalized appropriately [87,88].

3.2 Time-Averaged Transmission Function and Charge Transport through Linear Chains

To formulate a transport Hamiltonian, the coupling to left (L) and right (R) electrodes needs to be included. In a standard way, a tunnel Hamiltonian is used which will be treated later on within the wide band limit (see Section 2). The full Hamiltonian reads as follows [82]:

$$H = \sum_i \epsilon_i b_i^\dagger b_i + \sum_i V_{i,i+1} (b_i^\dagger b_{i+1} + \text{h.c.}) \qquad (18)$$

$$+ \sum_{k \in L} t_{k,L} (c_k^\dagger b_1 + \text{h.c.}) + \sum_{k \in R} t_{k,R} (c_k^\dagger b_N + \text{h.c.}) + \sum_{k \in L,R} \epsilon_k c_k^\dagger c_k.$$

Within the Landauer approach, the transmission function at a given instant of time t_i along the MD trajectory $T(E, t_i)$ for the previous model can be written as

$$T(E, t_i) = 4\gamma_L \gamma_R |G_{1N}(E, t_i)|^2, \tag{19}$$

where γ_L and γ_R are effective coupling terms to the L and R electrodes within the wide-band approximation, respectively. $G_{1N}(E, t_i)$ is the $1, N$-matrix element of the chain Green's function, which can be calculated via a matrix Dyson equation:

$$
\begin{aligned}
\mathbf{G}^{-1}(E, t_i) &= E\mathbf{1} - \mathbf{H}(t_i) - \Sigma_L - \Sigma_R, \\
(\Sigma_L)_{lj} &= -i\,\gamma_L \delta_{l1} \delta_{j1}, \\
(\Sigma_R)_{lj} &= -i\,\gamma_R \delta_{lN} \delta_{jN}.
\end{aligned}
\tag{20}
$$

Notice that the above expressions refer to transport characteristics at a given time t_i along the MD trajectory. Thus, a time series of transmission functions will be generated which is then averaged over the simulation time.

3.3 Molecular Dynamics Methodology

There are a variety of different standard software to perform classical MD simulations of biological systems. In the approach described here, the AMBER-parm99 force field [99] with the parmBSC0 extension [100] as implemented in the GROMACS [101] software package was employed.

After a standard heating procedure followed by a 1 ns of equilibration phase which is discarded afterwards, a 30 ns MD run with a time step of 2 fs was used. The simulations were carried out in a rectangular box with periodic boundary conditions and filled with 5500 TIP3P [102] water molecules and 20 sodium counterions for neutralization. Snapshots of the molecular structures were saved every 1 ps, for which the charge transfer parameters were calculated with the SCC-DFTB-FO approach as described above. To assess the effect of environment, these parameters were computed with and without the external charges in Eq. (17).

3.4 Charge Transport and Dynamics in Short DNA Sequences

Figure 7(a) shows the calculated transmission for various ideal (static) B-DNA sequences. As expected, the transmission spectrum consists of a set of resonances which can be related to the eigenvalues of the corresponding

a) ideal static B-DNA

b) MD without solvent effects

c) MD including solvent effects

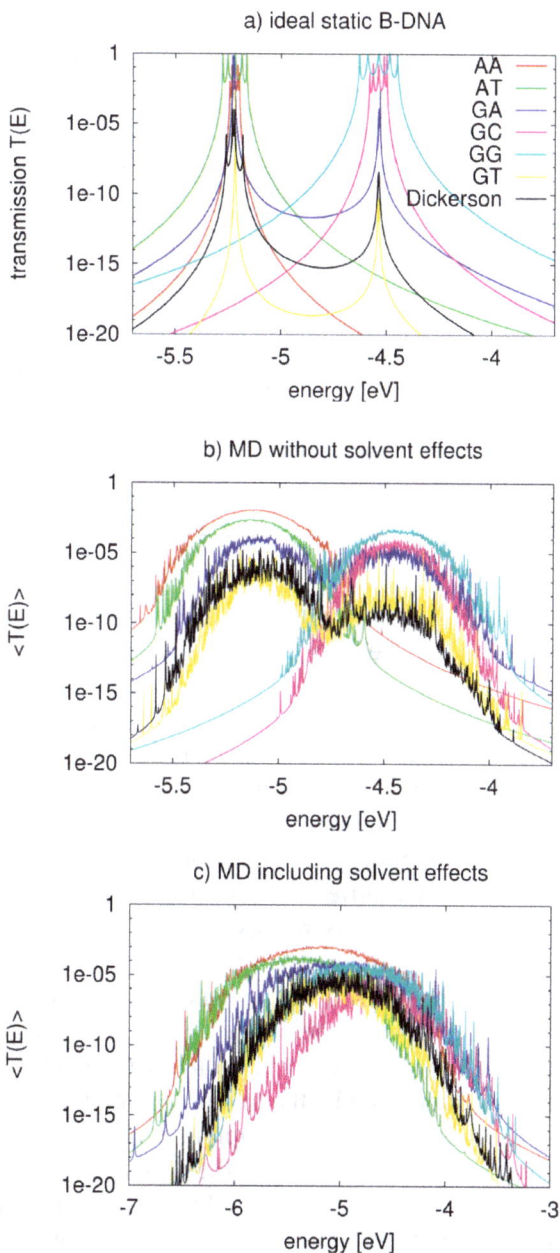

Figure 7 Transmission of the ideal chain (a) including dynamical effects (b) and the effect of environment (c) for various DNA sequences. Note the broader energy range in (c). Reprinted with permission from [82]. © 2009 American Institute of Physics.

Table 1 Electronic couplings V_{ij} for hole transfer in idealized static A and B-DNA without QM/MM environemnt compared to MD averaged values with standard deviations $\langle V_{ij}\rangle \pm \sigma$ including the QM/MM environment, for helical parameters of the idealized A and B-DNA, see [103] and [104], all values in eV. Reprinted with permission from [82]. © 2009 American Institute of Physics.

	static B-DNA		average MD values		static A-DNA	
XY	5'-XY-3'	5'-YX-3'	5'-YX-3'	5'-XY-3'	5'-XY-3'	5'-YX-3'
	V_{ij}	V_{ij}	$\langle V_{ij}\rangle \pm \sigma$	$\langle V_{ij}\rangle \pm \sigma$	V_{ij}	V_{ij}
intrastrand						
AA	0.013		0.058 ± 0.037		0.070	
GG	0.052		0.029 ± 0.023		0.012	
GA	0.053	0.026	0.034 ± 0.027	0.033 ± 0.028	0.023	0.044
interstrand						
GC	0.017	0.029	0.012 ± 0.012	0.022 ± 0.016	0.006	0.054
AT	0.035	0.031	0.037 ± 0.029	0.045 ± 0.034	0.018	0.107
GT	0.020	0.005	0.016 ± 0.013	0.026 ± 0.023	0.010	0.073

Hamiltonian matrices. Due to the very small values of the electronic coupling parameters, these resonances lie very close to the onsite energies of the respective fragments. Also expected are the reduced values of the transmission for inhomogeneous sequences due to the increased energy gaps arising from the differences in the ionization potentials from base to base.

In a second step the coupling parameters are now evaluated along the MD trajectories but omitting the QM/MM term (Q_A) in Eq. (17). Table 1 shows the MD-averaged couplings in comparison to those of the ideal A- and B-DNA structures. Notice the differences in the coupling when comparing with the static conformations. This results strongly suggests the averaged MD structures being significantly different from the ideal ones [88]. The role of fluctuations is further reflected in the variances σ which are of the same order of magnitude as the averages themselves. These results are nearly independent of the interaction with solvent, indicating that the electronic coupling fluctuations are mainly dominated by the mutual orientation of the base pairs and are not sensitive to the electrostatic coupling to the environment [88].

Figure 7(b) shows the transmission for the various sequences including only the internal dynamics. As a result of the broad distribution of the onsite energies, the transmission spectrum broadens. Further, the dynamical disorder of onsite energies increases the transmission of low-conducting (static) structures, while it decreases it for the "high-conducting" ones. This can be understood by taking into account that the fluctuations of the onsite energies lead to conformations for the "mixed" sequences, such as poly(GA), poly(GT), and the Dickerson dodecamer, where the effective energy gaps become smaller than in the idealized static structures. Therefore, CT-active conformations arise due to the dynamics. On the contrary, the homogeneous sequences become effectively disordered due to the dynamical fluctuations thus reducing the transmission.

To include the effect of DNA backbone, water and counter-ions, the Hamiltonian in Eq. (17) is used to calculate the new electronic coupling and onsite energy parameters. The electric field induced by the water molecules leads to large fluctuations of the onsite energies in the order of 0.4 eV compared to only 0.14 eV without the environment. Also the averages, $\langle \epsilon_i \rangle$ are shifted by 0.2–0.3 eV to lower energies. Figure 7(c) shows the transmission of the DNA species in the presence of electrostatic field induced by the environment. Since the environment does not affect the electronic coupling strongly, the main difference from Figure 7(b) arises from the larger fluctuations of onsite energy values. As a result of the wider distribution of onsite energies as well as the environment-induced energy shifts, the transmission spectra become considerably broader.

Two interesting points can also be seen in Figure 7(c). First, poly(A) shows the largest transmission, in contrast to the static case where poly(G) is better conducting. Second, the transmission of the heterogeneous species like poly(GA), poly(GT) and the Dickerson dodecamer sequence increases substantially compared to the idealized, static case, indicating that the fluctuations of onsite energies may lead to conformations with smaller effective onsite-disorder.

3.5 Correlations Matter

Several theoretical studies of DNA conduction have used static disorder to address the influence of the solvent or of inhomogeneous base sequences [42, 43, 45]. However, the question arises whether temporal (dynamical) correlations are important in determining the charge transfer efficiency. Taking as a en example the case of a poly(A) heptamer, this issue has been analyzed in some detail. To progressively increase the degree of correlations, three different cases for the probability distribution of the site energies have been chosen: (i) the Anderson model [105], where the onsite energies are randomly drawn from a square-box distribution of width w with uniform probability $P(\epsilon) = 1/2w$. The box width is $w = \sqrt{3}\sigma$, where $\sigma(\epsilon)$ is the standard deviation of onsite energies resulting from the MD simulations ($\sigma \sim 0.4$ eV);

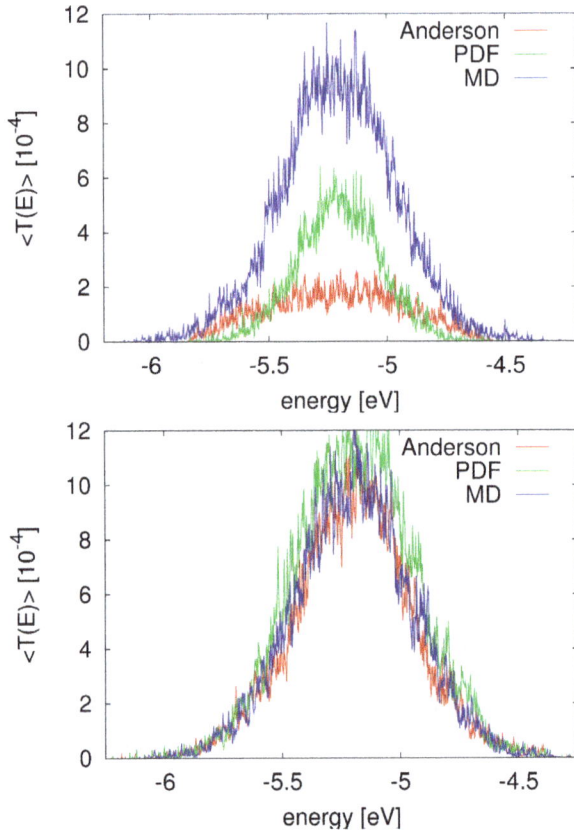

Figure 8 Comparison of $\langle T(E) \rangle$ for the MD simulation of a poly(A) heptamer with two statistical models. Top panel: The average transmission function is calculated for onsite energies from the MD simulations time series (blue); for onsite energies drawn from the respective probability distribution functions on each site (green); and the Anderson model (red) where all onsite energies are randomly drawn from a square-box distribution. Bottom panel: the original MD time series of onsite energies is used, the same for the three models, while $\langle T(E) \rangle$ is calculated for electronic couplings V_{ij} from the original MD time series (blue); for V_{ij} drawn from their respective probability distribution functions (green); and the Anderson model (red), respectively. Reprinted with permission from [82]. © 2009 American Institute of Physics.

(ii) a PDF model where the onsite energies are drawn randomly from a normal distribution $P(\epsilon_i)$ at each site i of the chain, but with no inter-site correlations. These distributions were however obtained from MD simulations; (iii) the full time series are used encoding all relevant time correlations.

To make the results comparable, the electronic couplings are taken constant $V_{ij} = 0.05$ eV. The average transmission of the poly(A) heptamer for the three cases is shown in the top panel of Figure 8. Clearly, the Anderson model largely suppresses transport (low average transmission) hinting at the potential relevance of correlated fluctuations. Neglecting non-local correlations as in the PDF model but still using a more realistic distribution function of the site energies leads to an increase of the transmission probability. The maximal transmission is obtained when including the full time series in the calculation, which is a clear indicator that non-local (site-to-site) correlations are crucial in determining the charge transport efficiency. Notice that the influence of correlations in the hopping integrals does not seem to be as dramatic as for the onsite energies (bottom panel of Figure 8).

3.6 Conformational Analysis

To analyze in more detail the role played by the conformational dynamical disorder, two effective measures can be introduced [82]:

$$\Sigma = \sqrt{\frac{1}{N} \sum_{i=1}^{N} (\epsilon_i - \langle \epsilon \rangle_N)^2} = \sqrt{\langle \epsilon^2 \rangle_N - \langle \epsilon \rangle_N^2} \tag{21}$$

$$\Pi = \prod_{i=1}^{N-1} V_{i,i+1} \tag{22}$$

The standard deviation Σ is calculated for the ϵ_i along the chain and has an evident meaning. Large values of Σ indicate large differences of neighboring site-energies. Note, the index N in $\langle \epsilon \rangle_N$ and $\langle \epsilon^2 \rangle_N$ means that averaging is performed for the N sites along the chain. The other parameter Π is motivated by the form of Greens function matrix element $G_{1N}(E)$ required to calculate the transmission function, which scales approximately as the product of electronic couplings in the weak coupling case, i.e. when the ratio $V/\Delta\epsilon \ll 1$, where V and $\Delta\epsilon$ are typical hopping matrix elements and energy gap parameters, respectively. Thus, this quantity determines the transmission efficiency of the system; small values of Π account for conformations with small couplings along the DNA chain. In order to reduce the complexity of further analysis we additionally define the value $T_{\max}$ as simply the maximum of a given transmission function $T(E)$. Note, that the value $T_{\max}$ can be located anywhere within the respective energy range. All three parameters

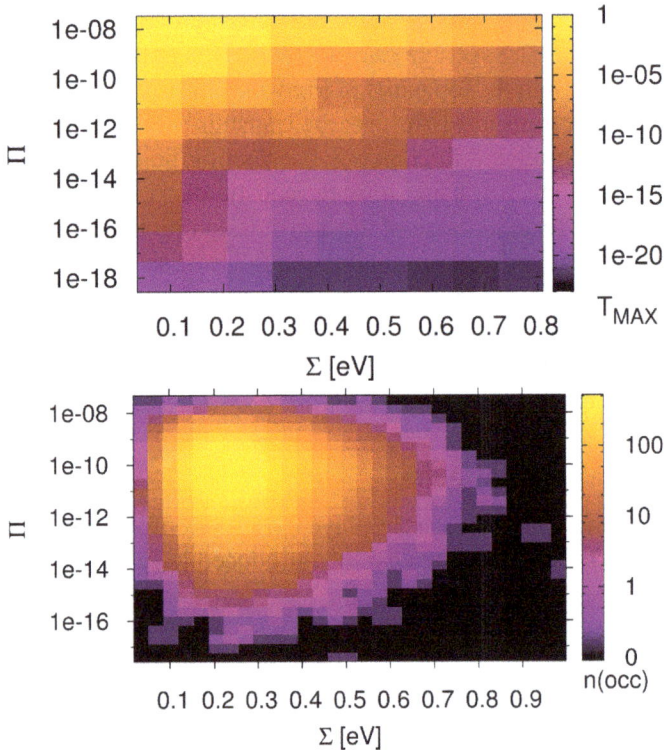

Figure 9 Statistical analysis of T_{max} in a poly(G) heptamer, for the 30 ns data with electronic parameters for every ps (30,000 DNA conformations). T_{max} depending on Σ and Π (top); number of conformations found in a given interval of Σ and Π (bottom). Reprinted with permission from [82]. © 2009 American Institute of Physics.

Σ, Π and T_{max} are now calculated for 30,000 snapshots along the 30 ns MD trajectory of a poly(G) heptamer.

The results are shown in Figure 9 (top panel). We see that none of the measures Σ and Π alone is able to describe the conformations of high conduction, but both seem to contribute nearly linearly to the transmission (note the logarithmic scales for Π and T_{max}). However, for the transport active conformations, small Σ and large Π values are required.

Figure 9 (top panel) also shows that T_{MAX} depends more strongly on Π than on Σ. For instance, if Π is kept fixed at 10^{-8}, then the maximum transmission T_{MAX} is still at least 10^{-7} for all values of Σ. On the other hand, keeping the parameter Σ fixed makes T_{max} decrease to almost 10^{-17} even for the smallest value of Σ. The bottom panel of Figure 9 shows the corresponding occupation plot. Here, it is quantified how many conformations exhibit a certain combination of Σ and Π parameters. It seems that the number of transport active conformations with appropriate electronic couplings and onsite energies is very small. Most conformations have Π values of about 10^{-10} and Σ values of about 0.25 eV, and are therefore "CT-silent". This analysis is obviously subjected to the restriction that transport characteristics have been calculated using (time averaged) transmission functions, which eventually cannot catch the full transport pathways of the system (decoherence effects are clearly not included here). Nevertheless, this approach is able to shed light onto the concept of CT active or silent conformations.

4 Charge Transport in Dissipative Environments

In the previous section charge transport was treated within Landauer theory and the influence of dynamical fluctuations was effectively included via a time average procedure. Here, we will change the perspective and go back to model Hamiltonian formulations where the coupling to dynamical degrees of freedom is explicitly included. However, with the methodology presented in the preceding section we will now be in the position to parametrize not only the electronic structure part of the model but also the interaction with the vibrational system. As a representative example, we focus on the Dickerson dodecamer with the sequence $3'-\mathrm{GCGCTTAACGGC}-5'$ and for which the effect of the dynamical fluctuations becomes very clear (see Section 3, especially Figure 7). As introduced before, the starting point is a time-dependent electronic Hamiltonian for a linear chain where both onsite energies $\epsilon_j(t)$ and electronic coupling terms $V_{j,j+1}(t)$ are drawn from the MD simulations:

$$H = \sum_j \epsilon_j(t) b_j^\dagger b_j + \sum_j V_{j,j+1}(t)\, (b_j^\dagger b_{j+1} + \mathrm{h.c.}). \qquad (23)$$

Since Eq. (23) contains random variables through the time series. We are, strictly speaking, confronted with the problem of dealing with charge transport in an stochastic Hamiltonian. This is a complex task which has been addressed, e.g., in the context of exciton transport [106–109], but also to some degree in electron transfer theories [110–112]. Here, we adopt a different point of view and reformulate this model in a way that the coupling to dynamical degrees of freedom is split off and included in a bosonic bath, which can thus be explicitly treated. The Hamiltonian can be rewritten in the following way [83]:

$$H = \sum_j \langle \epsilon_j \rangle_t \, b_j^\dagger b_j - \sum_j \langle V_{j,j+1} \rangle_t \, (b_j^\dagger b_{j+1} + \text{h.c.})$$

$$+ \; H_{\text{bath}} + H_{\text{el-bath}} + H_{\text{tunnel}} + H_{\text{leads}} \qquad (24)$$

where

$$H_{\text{bath}} = \sum_\alpha \Omega_\alpha B_\alpha^\dagger B_\alpha$$

$$H_{\text{el-bath}} = \sum_{\alpha,j} \lambda_\alpha b_j^\dagger b_j (B_\alpha + B_\alpha^\dagger)$$

$$H_{\text{tunnel}} = \sum_{\mathbf{k},s,j} (t_{\mathbf{k}s,j} c_{\mathbf{k}s}^\dagger b_j + \text{h.c.})$$

$$H_{\text{leads}} = \sum_{\mathbf{k},s} \epsilon_{\mathbf{k}s} c_{\mathbf{k}s}^\dagger c_{\mathbf{k}s}$$

The time averages (over the corresponding time series) of the electronic parameters $\langle \epsilon_j \rangle_t$ and $\langle V_{j,j+1} \rangle_t$ have been split off to provide a zero-order electronic Hamiltonian which contains dynamical effects on a mean-field-like level. The effect of the fluctuations around these averages is hidden in the vibrational bath, which is assumed to be a collection of a large $(N \to \infty)$ number of harmonic oscillators in thermal equilibrium at temperature $k_B T$. The bath will be characterized by a spectral density $J(\omega)$ which can also be extracted from the MD simulations. The model is completed by including the interaction with electrodes along the same lines as in Eqs. (1) and (19).

The previous model relies on some basic assumptions that can be substantiated by the results of the MD simulations: (i) The complex DNA dynamics can be well mimic within the harmonic approximation by using a continuous vibrational spectrum; (ii) the simulations show that local onsite energy fluctuations are much stronger in presence of a solvent than those of the electronic hopping integrals, so that we assume that the bath is coupled only diagonally to the charge density fluctuations; and (iii) fluctuations on different sites display rather similar statistical properties, so that the charge-bath coupling λ_α is taken to be independent of the site j. A typical correlation function of the onsite energy fluctuations is displayed in Figure 10 for the Dickerson dodecamer in both solvent and vacuum (no electrostatic environment) conditions as well as one case of off-diagonal correlations between nearest-neighbor site energies (inset). Also shown are fits to stretched exponentials, which are in general a compact way of representing a fit to a sum of single exponential functions, i.e. the presence of different time scales in the problem, leading to long time tails in the correlation functions. As extensively discussed in, e.g., [113], the emergence of long-time tails can be generally understood in terms of the ratio between typical charge propagation times and typical time scales for the dynamical fluctuations of the system. Especially, long fluctuation time scales (compared with typical charge propagation times) can induce deviations from a purely single exponential behavior. From the figure it becomes clear that off-diagonal correlations decay on shorter time scales as the local ones, so that on a first approximation to neglect them can be justified (although in general they can not be fully excluded); further, the decay of the correlations for the vacuum simulations is considerably much faster than in a solvent indicating the strong influence of the latter in gating the electronic structure of the biomolecule.

Similar to Section 2, we perform a polaron transformation of the Hamiltonian Eq. (24), using the generator

$$\mathcal{U} = \exp\left[\sum_{\ell,\alpha} g_\alpha d_\ell^\dagger d_\ell (B_\alpha^\dagger - B_\alpha)\right].$$

The parameter $g_\alpha = \lambda_\alpha / \Omega_\alpha$ gives an effective measure of the electron-vibron coupling strength. As a result, we obtain a Hamiltonian with decoupled electronic and vibronic parts and where the onsite energies are shifted as

$$\langle \epsilon_j \rangle_t \rightarrow \langle \epsilon_j \rangle_t - \int_0^\infty d\omega J(\omega)/\omega.$$

The retarded Green function of the system is now an entangled electronic-vibronic object that can not be treated exactly; we thus decouple it in the approximate way, see also Eq. (6) [79, 114]:

$$
\begin{aligned}
\mathcal{G}_{nm}(t,t') &= -i\,\theta(t-t')\left\langle [d_n(t)\mathcal{X}^\dagger(t), d_m^\dagger(t')\mathcal{X}(t')]_+\right\rangle \qquad (25)\\
&\approx -i\,\theta(t-t')\left\{\langle d_n(t)d_m^\dagger(t')\rangle \langle \mathcal{X}^\dagger(t)\mathcal{X}(t')\rangle \right.\\
&\qquad \left. + \langle d_m^\dagger(t')d_n(t)\rangle \langle \mathcal{X}(t')\mathcal{X}^\dagger(t)\rangle\right\}\\
&= \theta(t-t')\left\{G_{nm}^>(t,t')e^{-\phi(t-t')} - G_{nm}^<(t,t')e^{-\phi(t'-t)}\right\}
\end{aligned}
$$

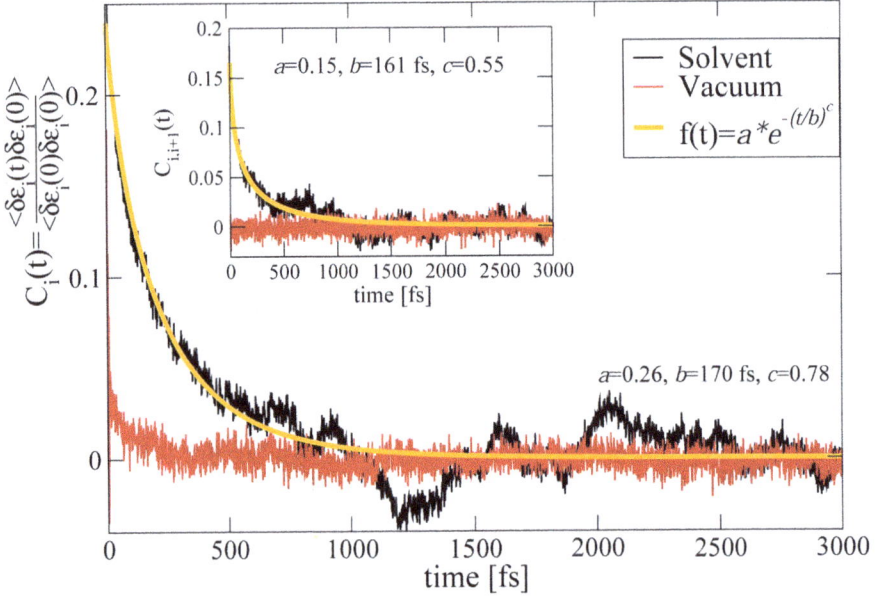

Figure 10 Normalized auto-correlation function $C_i(t) = \langle \delta\epsilon_i(t)\delta\epsilon_i(0)\rangle/\langle\delta\epsilon_i^2\rangle$ and averaged nearest-neighbor correlation function $C_{i,i+1}(t) = \langle\delta\epsilon_i(t)\delta\epsilon_{i+1}(0)\rangle/\langle\delta\epsilon_i\delta\epsilon_{i+1}\rangle$ (inset) of the onsite energy fluctuations. The solid lines are fits to stretched exponentials. Reprinted with permission from [83]. © 2010 Institute of Physics.

$$\phi(t) = \sum_\alpha \left(\frac{\lambda_\alpha}{\Omega_\alpha}\right)^2 \left[(1+N_\alpha)e^{-i\Omega_\alpha t} + N_\alpha e^{+i\Omega_\alpha t}\right]$$

In this equation, $\theta(t-t')$ is the Heaviside function and the pure bosonic operator $\mathcal{X}(t) = \exp[\sum_\alpha g_\alpha(B_\alpha^\dagger - B_\alpha)]$. In the last row of Eq. (25) we can pass to the continuum limit and express $\phi(t)$ in terms of the bath spectral density $J(\omega)$ [115]:

$$\phi(t) = \frac{1}{\hbar}\int_0^\infty d\omega \frac{J(\omega)}{\omega^2}\coth\frac{\hbar\omega}{k_B T}(1-\cos\omega t)$$

$$-i\frac{1}{\hbar}\int_0^\infty d\omega \frac{J(\omega)}{\omega^2}\sin\omega t. \tag{26}$$

Using standard techniques [83, 94], we can then write an expression for the electrical current:

$$I_L = \frac{2e}{\hbar} \int \frac{dE}{2\pi} \int \frac{dE'}{2\pi} T(E')$$

$$\times \left\{ f_L(E)(1 - f_R(E'))\Phi(E - E') - (1 - f_L(E))f_R(E')\Phi(E' - E) \right\},$$

$$\Phi(E) = \int \frac{dt}{\hbar} e^{\frac{i}{\hbar} E t} e^{-\phi(t)}. \tag{27}$$

Here, the transmission-like function $T(E)$ is given by $t(E) = \mathrm{Tr}\{\mathbf{G}_0(E)\mathbf{\Gamma}_L\mathbf{G}_0^\dagger(E)\mathbf{\Gamma}_R\}$ and is calculated without including the coupling to the bosonic bath, which is already taken into account by the $\Phi(E)$ functions.

4.1 Getting the Bath Spectral Density from Molecular Dynamics

A central issue in the reformulation of the transport problem is how to get the bath spectral density $J(\omega)$ from the information encoded in the time series of the electronic parameters. We will explain this point by using a much simpler toy model consisting of a single time dependent level whose site energy is a Gaussian random variable:

$$H = \delta\epsilon(t)b^\dagger b + H_{\text{tunnel}} + H_{\text{leads}} \tag{28}$$

Using equation of motion techniques for the Green function $G(t, t') = -(i/\hbar)\theta(t - t')\langle\{b(t), b^\dagger(t')\}\rangle$ of the system we arrive at the following solution (within the wide band approximation in the coupling to the electrodes:

$$G(t, t') = -\frac{i}{\hbar}\theta(t - t')\mathcal{U}(t, t'), \tag{29}$$

where $\mathcal{U}(t, t') = \exp(-(i/\hbar)\int_{t'}^t ds(\delta\epsilon(t) - i\Gamma))$. Averaging the Green function over the distribution of the random variable $\delta\epsilon(t)$ and performing a cumulant expansion up to second order (Gaussian distribution) yields now in the energy-space:

$$\langle G(E)\rangle = -\frac{i}{\hbar}\int_0^\infty dt\, e^{\frac{i}{\hbar}(E + i\Gamma)t}\, e^{-\frac{1}{\hbar^2}\int_0^t ds \int_0^s ds'\, \langle\delta\epsilon(s)\delta\epsilon(s')\rangle}, \tag{30}$$

which is the formally exact solution of the problem if the correlation function $\langle\delta\epsilon(s)\delta\epsilon(s')\rangle$ is specified. We may now look at the same problem from a different point of view by considering the coupling of a single site with *time-independent* onsite energy to a continuum of vibrational excitations. Along similar lines as in the previous section, we can write the retarded Green function as

$$G(E) = -\frac{i}{\hbar} \int_0^\infty dt \, e^{\frac{i}{\hbar}(E+i\Gamma)t} e^{-\phi(t)}, \tag{31}$$

where $\phi(t)$ has been already defined in Eq. (26). By comparison of Eqs.(30) and (31), it becomes clear that there should exist a relation between the (real) correlation function and the real part of $\phi(t)$. Writing $\mathrm{Re}\,\phi(t)$ as

$$\mathrm{Re}\,\phi(t) = \frac{1}{\hbar} \int_0^\infty d\omega \, \frac{J(\omega)}{\omega^2} \coth \frac{\hbar\omega}{k_B T} (1 - \cos \omega t)$$

$$= \int_0^t ds \int_0^s ds' \left\{ \frac{1}{\hbar} \int_0^\infty d\omega \, J(\omega) \coth \frac{\hbar\omega}{k_B T} \cos [\omega \, (s - s')] \right\},$$

we can conclude that

$$\langle \delta\epsilon(s)\delta\epsilon(s') \rangle = \hbar \int_0^\infty d\omega \, J(\omega) \coth \frac{\hbar\omega}{k_B T} \cos [\omega \, (s - s')].$$

Upon inversion, we get

$$J(\omega) = \frac{2}{\pi\hbar} \tanh \frac{\hbar\omega}{k_B T} \int_0^\infty dt \, \cos \omega t \, C(t) = \frac{2}{\pi\hbar} \tanh \frac{\hbar\omega}{k_B T} j(\omega), \tag{32}$$

which provides the desired relation between the bath spectral properties and the correlation function of the onsite energy fluctuations.

4.2 Charge Transport in the Dickerson Dodecamer

We have applied this formulation to study charge transport through the Dickerson dodecamer in presence and absence of solvent effects. To first illustrate the influence of the solvent, in Figure 11 the time averaged onsite energies along the chain are displayed. Remarkably, the presence of the solvent "smoothes" the averaged energy profile (though the amplitude of the fluctuations clearly becomes stronger).

In Figure 12 the current calculated with Eq. (27) is shown for the two cases of interest. Due to the presence of tunnel barriers in the wire which, on average, are not fully compensated by the gating effect of the environment, the absolute current values are rather small when compared with those of homogeneous sequences (not shown) [81]. However, the current including the solvent is roughly fifteen times larger than for that obtained from the

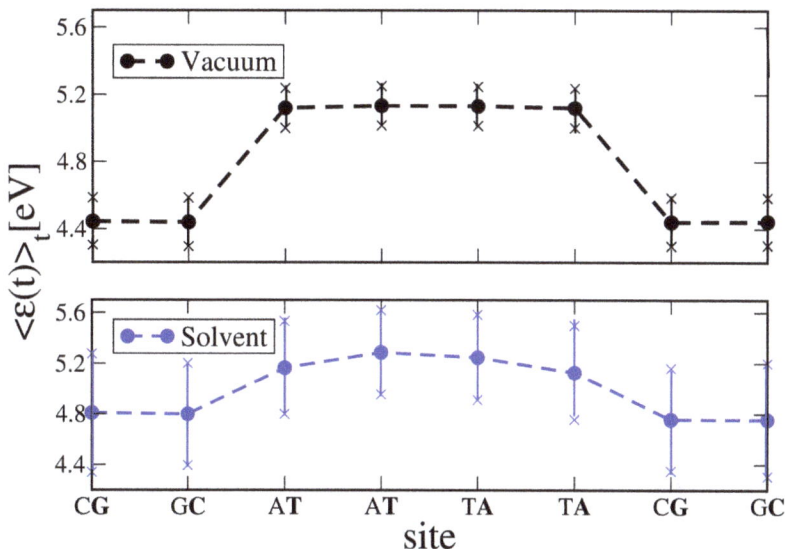

Figure 11 Time average of the absolute value of the onsite energies of the Dickerson dode-camer for simulations in vacuum (upper panel) and in solvent (lower panel). Notice that the averaged energy profile in presence of the solvent becomes smoother but also the fluctuations around the averages are stronger. This smoothing reduces the energy barriers between the sites and hence favors charge migration. Reprinted with permission from [83]. © 2010 Institute of Physics.

simulations in vacuum. Although our model Hamiltonian in Eq. (25) does not fully contain all the dynamical correlations encoded in the time dependent electronic parameters, we nevertheless expect that their inclusion would lead to an even further increase of the difference between solvent and vacuum results. This nicely demonstrates that coupling to dynamical degrees of freedom is crucial when dealing with charge migration in soft systems and that the basic effects can still be catch by effective model Hamiltonians whenever appropriate (realistic) parameterizations are carried out. The possibility to parameterize the bath spectral density $J(\omega)$ using the information obtained from the MD time series makes our approach very efficient. We remark that the scheme presented here is obviously not limited to the treatment of DNA but it can equally well be applied to deal with charge migration in other complex systems like molecular organic crystals or polymers, where charge dynamics and coupling to fluctuating environments plays an important role [110–112, 116–120].

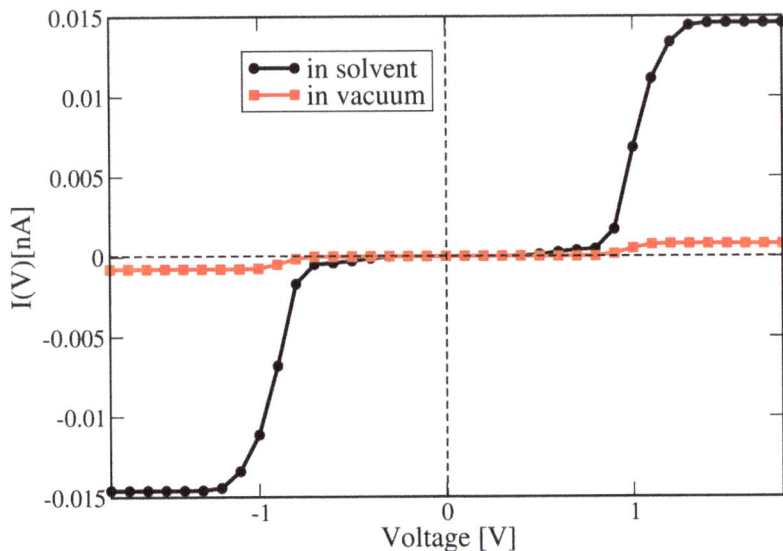

Figure 12 Electrical current for the Dickerson dodecamer in both vacuum (no solvent) and including the solvent. The current is considerably enhanced upon inclusion of solvent fluctuations, nicely demonstrating the strong gating effect of the latter onto the energy profile of the DNA chain. Reprinted with permission from [83]. © 2010 Institute of Physics.

5 Conclusions

In contrast to hole transfer in solution, where there seems to be common agreement about the microscopic charge migration mechanism (incoherent hopping processes), the clarification of the most relevant charge transport pathways in short DNA oligomers still remains elusive. This may be due to the fact, on one side, that well-controlled electrical transport experiments are difficult to perform with biomolecules, so that general trends and dependencies on base sequences, temperature, etc. have not been fully elucidated. On the other side, the intrinsic complexity of biomolecules make a description in terms of simple models very challenging and as a consequence, many results can be questionable if the physically relevant parameter region is not well defined. The fundamental role played by the structural dynamics and the need to include such effects in a non-perturbative way in charge transport calculations further increases the problems faced by theoreticians when dealing with the biomolecular electrical response. In this paper, we have focused on recently developed methodologies which try to address all these previously mentioned problems by combining molecular dynamics simulations with electronic structure calculations to parameterize effective model Hamiltonians able to deal with different transport scenarios in DNA oligomers. The main advantage of this approach is the possibility to develop a controlled coarse-graining of the electronic structure, which in its turn allows tuning the degree of complexity of the model Hamiltonians for transport. Additionally, the methodology is flexible enough to be transferred to the study of charge transport and dynamics in other systems like polymers or organic crystals, thus providing a common platform to treat with such issues. Obviously, several points remain still open and should be addressed in the future. We only mention two of them: How reliable is a quantum mechanically computed electronic structure along a *classical* molecular dynamic trajectory, i.e., with geometries obtained using classical interaction potentials? How strong is the influence of a charge injected into the system onto the underlying electronic structure? This latter effect is usually neglected in transport calculations, but it has been recently shown that it has an influence in the prediction of, e.g., the onset of diffusive transport in organic stacks [121], so that we may expect it to play also a critical role in determining the charge migration efficiency in biomolecular systems.

Acknowledgments

The work presented in Section 2 is the result of a fruitful cooperation with Marcus Elstner, Tomas Kubar, and Benjamin Woiczikowski. We further acknowledge very fruitful discussions with Stanislav Avdoshenko, Yuri Berlin, Giorgia Brancolini, Rosa Di Felice, Joshua Jortner, Myeong Lee, Pedro Manrique, Ron Naaman, Danny Porath, Stephan Roche, Dmitri Ryndyk, Jewgeni Starikov, and Wei Yuan Tu.

This work has been supported by the Volkswagen Foundation grant No. I/78-340, by the EU under contract IST-2001-38951, by the Deutsche Forschungsgemeinschaft under contracts CU 44/5-2, CU 44/5-3, and CU 44/3-2 as well as by the South Korea Ministry of Education, Science and Technology Program "World Class University" under contract R31-2008-000-10100-0. We also acknowledge the Center for Information Services and High Performance Computing (ZIH) at the Dresden University of Technology for computational resources. We further gratefully acknowledge support from the German Excellence Initiative via the Cluster of Excellence EXC 1056 "Center for Advancing Electronics Dresden" (cfAED).

References

[1] J. R. Heath, M. A. Ratner, Physics Today (May 2003).

[2] A. Aviram, M. A. Ratner, Annals of the New York Academy of Sciences, Vol. 852 (The New York Academy of Sciences, New York, 1998).

[3] A. Aviram, M. A. Ratner, V. Mujica, Annals of the New York Academy of Sciences, Vol. 960 (The New York Academy of Sciences, New York, 2002).

[4] G. Cuniberti, G. Fagas, K. Richter, Introducing Molecular Electronics, Lecture Notes in Physics, Vol. 680 (Springer, Berlin, 2005).

[5] K. Keren, R. S. Berman, E. Buchstab, U. Sivan, E. Braun, Science, 302, 1380–1382 (2003).

[6] E. Braun, Y. Eichen, U. Sivan, G. Ben-Yoseph, Nature, 391, 775–778 (1998).

[7] M. Mertig, R. Kirsch, W. Pompe, H. Engelhardt, European Physics Journal D, 9, 45–48 (1999).

[8] C. R. Treadway, M. G. Hill, J. K. Barton, Chemical Physics, 281, 409–428 (2002).

[9] C. J. Murphy, M. R. Arkin, Y. Jenkins, N. D. Ghatlia, S. H. Bossmann, N. J. Turro, J. K. Barton, Science, 262, 1025–1029 (1993).

[10] N. J. Turro, J. K. Barton, Journal of Biological and Inorganic Chemistry, 3, 201–209 (1998).

[11] E. M. Boon, J. K. Barton, Current Opinions in Structural Biology, 12, 320–329 (2002).

[12] C. Wan, T. Fiebig, S. O. Kelley, C. R. Treadway, J. K. Barton, Proceedings of the National Academy of Sciences USA, 96, 6014–6019 (1999).

[13] S. O. Kelley, N. M. Jackson, M. G. Hill, J. K. Barton, Angewandte Chemie International Edition, 38, 941–945 (1999).

[14] M. Bixon, B. Giese, S. Wessely, T. Langenbacher, M. E. Michel-Beyerle, J. Jortner, Proceedings of the National Academy of Sciences USA, 96, 11713–11716 (1999).

[15] E. Meggers, M. E. Michel-Beyerle, B. Giese, Journal of the American Chemical Society, 120, 12950–12955 (1998).

[16] F. D. Lewis, X. Liu, Y. Wu, S. E. Miller, M. R. Wasielewski, R. L. Letsinger, R. Sanishvili, A. Joachimiak, V. Tereshko, M. Egli, Journal of the American Chemical Society, 121, 9905–9906 (1999).

[17] D. Ly, L. Sanii, G. B. Schuster, Journal of the American Chemical Society, 121, 9400–9410 (1999).

[18] P. T. Henderson, G. Hampikian, D. Jones, Y. Kan, G. B. Schuster, Proceedings of the National Academy of Sciences USA, 96, 8353–8358 (1999).

[19] G. B. Schuster, Topics in Current Chemistry, Vol. 237 (Springer, Berlin, 2004).

[20] J. Jortner, M. Bixon, T. Langenbacher, M. E. Michel-Beyerle, Proceedings of the National Academy of Sciences USA, 95, 12759–12765 (1998).

[21] D. Porath, A. Bezryadin, S. D. Vries, C. Dekker, Nature, 403, 635–638 (2000).

[22] C. Nogues, S. R. Cohen, S. S. Daube, R. Naaman, Physical Chemistry Chemical Physics, 6, 4459–4466 (2004).

[23] A. J. Storm, J. V. Noort, S. D. Vries, C. Dekker, Applied Physics Letters, 79, 3881–3883 (2001).

[24] K.-H. Yoo, D. H. Ha, J.-O. Lee, J. W. Park, Jinhee Kim, J. J. Kim, H.-Y. Lee, T. Kawai, H. Y. Choi, Physical Review Letters, 87, 198102–198105 (2001).

[25] B. Xu, P. Zhang, X. Li, N. Tao, Nano Letters, 4, 1105–1108 (2004).

[26] H. Cohen, C. Nogues, R. Naaman, D. Porath, Proceedings of the National Academy of Sciences USA, 102, 11589–11593 (2005).

[27] X. Guo, A. A. Gorodetsky, J. Hone, J. K. Barton, C. Nuckolls, Nature Nanotechnology, 3, 163–167 (2008).

[28] J. Jortner, M. Bixon, T. Langenbacher, M. Michel-Beyerle, Proceedings of the National Academy of Sciences USA, 95, 12759–12765 (1998).

[29] A. Voityuk, N. Rösch, M. Bixon, J. Jortner, Journal of Physical Chemistry B, 104, 5661–5665 (2000).

[30] J. Jortner, M. Bixon, A. A. Voityuk, N. Rösch, Journal of Physical Chemistry A, 106, 7599–7606 (2002).

[31] Y. A. Berlin, A. L. Burin, M. A. Ratner, Journal of Physical Chemistry A, 104, 443–445 (2000).

[32] Y. A. Berlin, A. L. Burin, M. A. Ratner, Journal of the American Chemical Society, 123, 260–268 (2001).

[33] F. Grozema, S. Tonzani, Y. Berlin, G. Schatz, L. Sibbeles, M. Ratner, Journal of the American Chemical Society, 130, 5157–5166.(2008)

[34] Y. Berlin, A. L. Burin, M. A. Ratner, Chemical Physics, 275, 61–74 (2002).

[35] D. M. Basko, E. M. Cornwell, Physical Review Letters, 88, 098102–098105 (2002).

[36] E. M. Conwell, Proceedings of the National Academy of Sciences USA, 102, 8795–8799 (2005).

[37] V. M. Kucherov, C. D. Kinz-Thompson, E. M. Conwell, Journal of Physical Chemistry C, 114, 1663–1666 (2010).

[38] S. M. Kravec, C. D. Kinz-Thompson, E. M. Conwell, Journal of Physical Chemistry B, 115, 6166–6171 (2011).

[39] S.-P. Liu, A. Erbe, S. H. Weisbrod, Z. Tang, A. Marx, E. Scheer, Angewandte Chemie International Edition, 49, 3313–3316 (2010).

[40] T. Chakraborty, Charge Migration in DNA: Perspectives from Physics, Chemistry and Biology (Springer, Berlin/New York, 2007).

[41] G. Cuniberti, L. Craco, D. Porath, C. Dekker, Physical Review B, 65, 241314–241317 (2002).

[42] S. Roche, Physical Review Letters, 91, 108101–108104 (2003).

[43] D. Klotsa, R. A. Roemer, M. S. Turner, Biophysical Journal, 89, 2187–2198 (2005).

[44] E. Macia, F. Triozon, S. Roche, Physical Review B, 71, 113106–113109 (2005).

[45] S. Roche, D. Bicout, E. Macia, E. Kats, Physical Review Letters, 91, 22810–22813 (2003).

[46] J. Yi, Physical Review B, 68, 193103–193106 (2003).

[47] H. Yamada, E. Starikov, D. Hennig, European Physics Journal B, 59, 185–192 (2007).

[48] G. Cuniberti, E. Macia, A. Rodriguez, and R. A. Romer, in Charge Migration in DNA: Perspectives from Physics, Chemistry and Biology, edited by T. Chakraborty (Springer, Berlin/New York, 2007).

[49] R. Gutierrez, G. Cuniberti, in NanoBioTechnology: BioInspired Device and Materials of the Future, edited by O. Shoseyov and I. Levy (Humana Press, 2007).

[50] E. Artacho, M. Machado, D. Sanchez-Portal, P. Ordejon, J. M. Soler, Molecular Physics, 101, 1587–1594 (2003).

[51] P. J. de Pablo et al., Physical Review Letters, 85, 4992–4995 (2000).

[52] A. Hübsch, R. G. Endres, D. L. Cox, R. R. P. Singh, Physical Review Letters, 94, 178102–178105 (2005).

[53] R. G. Endres, D. L. Cox, R. R. P. Singh, S. K. Pati, Physical Review Letters, 88, 166601–166604 (2002).

[54] A. C. H. Zhang, R. D. Felice, Journal of Physical Chemistry B, 109, 15345–15348 (2005).

[55] A. Calzolari, R. D. Felice, E. Molinari, A. Garbesi, Applied Physics Letters, 80, 3331–3333 (2002).

[56] R. D. Felice, A. Calzolari, E. Molinari, Physical Review B, 65, 045104–045113 (2002).

[57] R. D. Felice, A. Calzolari, H. Zhang, Nanotechnology, 15, 1256–1263. (2004).

[58] J. P. Lewis, P. Ordejon, O. F. Sankey, Physical Review B, 55, 6880–6887 (1997).

[59] F. L. Gervasio, P. Carolini, M. Parrinello, Physical Review Letters, 89, 108102–108105 (2002).

[60] H. Mehrez, M. P. Anantram, Physical Review B, 71, 115405–115409 (2005).

[61] R. N. Barnett, C. L. Cleveland, A. Joy, U. Landman, G. B. Schuster, Science, 294, 567–571 (2001).

[62] D. N. Beratan, S. Priyadarshy, S. M. Risser, Chemistry and Biology, 4, 3–8 (1997).

[63] I. A. Balabin, D. N. Beratan, S. S. Skourtis, Physical Review Letters, 101, 158102–158105 (2008).

[64] R. Bruinsma, G. Gruener, M. R. D'Orsogna, J. Rudnick, Physical Review Letters, 85, 4393–4396 (2000).

[65] Y. A. Berlin, A. L. Burin, L. D. A. Siebbeles, M. A. Ratner, Journal Physical Chemistry A, 105, 5666–5678 (2001).

[66] F. Grozema, Y. Berlin, L. D. A. Siebbeles, Journal of the American Chemical Society, 122, 10903–10909 (2000).

[67] S. S. Mallajosyula, J. C. Lin, D. L. Cox, S. K. Pati, R. R. P. Singh, Physical Review Letters, 101, 176805–176808 (2008).

[68] A. Troisi, G. Orlandi, Journal Physical Chemistry B, 106, 2093–2101 (2002).

[69] T. Cramer, S. Krapf, T. Koslowski, Journal Physical Chemistry C, 111, 8105–9113 (2007).

[70] F. C. Grozema, S. Tonzani, Y. A. Berlin, G. C. Schatz, L. D. Siebbeles, M. A. Ratner, Journal of the American Chemical Society, 130, 5157–5166 (2008).

[71] A. A. Voityuk, Journal of Chemical Physics, 128, 115101–115105 (2008).

[72] A. A. Voityuk, K. Siriwong, N. Rösch, Angewandte Chemie International Edition, 43, 624–627 (2004).

[73] S. Chaudhury, B. J. Cherayil, Journal of Chemical Physics, 127, 145103–145108 (2007).

[74] W. Min, X. S. Xie, Physical Review E, 73, 010902–010905 (2006).

[75] S. Wennmalm, L. Edman, R. Rigler, Chemical Physics, 247, 61–67 (1999).

[76] J. Kim, S. Doose, H. Neuweiler, M. Sauer, Nucleic Acids Research, 34, 2516–2527 (2006).

[77] R. Gutierrez, S. Mandal, G. Cuniberti, Nano Letters, 5, 1093–1097 (2005).

[78] R. Gutierrez, S. Mandal, G. Cuniberti, Physcial Review B, 71, 235116–235124 (2005).

[79] R. Gutierrez, S. Mohapatra, H. Cohen, D. Porath, G. Cuniberti, Physical Review B, 74, 235105–235114 (2006).

[80] B. B. Schmidt, M. H. Hettler, G. Schön, Physcial Review B, 75, 115125–115132 (2007).
[81] R. Gutierrez, R. Caetano, P. B. Woiczikowski, T. Kubar, M. Elstner, G. Cuniberti, Physcial Review Letters, 102, 208102–208105 (2009).
[82] P. Woiczikowski, T. Kubar, R. Gutierrez, R. Caetano, G. Cuniberti, M. Elstner, Journal of Chemical Physics, 130, 215104–215127 (2009).
[83] R. Gutierrez, R. Caetano, P. B. Woiczikowski, T. Kubar, M. Elstner, G. Cuniberti, New Journal of Physics, 12, 023022–023038 (2010).
[84] B. Popescu, P. B. Woiczikowski, M. Elstner, and U. Kleinekathöfer, Physical Review Letters, 109, 176802–176805 (2012).
[85] M. Lee, S. Avdoshenko, R. Gutierrez, G. Cuniberti, Physcial Review B, 82, 155455–155461 (2010).
[86] M. H. Lee, G. Brancolini, R. Gutierrez, R. Di Felice, G. Cuniberti, Journal of Physical Chemistry B, 116, 10977–10985 (2012)
[87] T. Kubar, P. B. Woiczikowski, G. Cuniberti, M. Elstner, Journal of Physical Chemistry B, 112, 7937–7947 (2008).
[88] T. Kubar, M. Elstner, Journal of Physical Chemistry B, 112, 8788–8798 (2008).
[89] H. Yamada, Physics Letters A, 332, 65–73 (2004).
[90] H. Yamada, E. B. Starikov, D. Hennig, J. F. R. Archilla, cond-mat/0406040 (2004).
[91] R. A. Caetano, P. A. Schulz, Physical Review Letters, 95, 126601–126604 (2005).
[92] X. F. Wang, T. Chakraborty, Physical Review Letters, 97, 106602–106605 (2006).
[93] G. D. Mahan, Many Particle Physics (Plenum Press, New York, 2000).
[94] Y. Meir, N. S. Wingreen, Physical Review Letters, 68, 2512–2515 (1992).
[95] J. Koch, F. von Oppen, Physical Review Letters, 94, 206804–206807 (2005).
[96] E. B. Starikov, Philosophical Magazine, 85, 3435–3452 (2005).
[97] T. Frauenheim, G. Seifert, M. Elstner, Z. Hajnal, et al., Physica Status Solidi (b), 217, 41–62 (2000).
[98] K. Senthilkumar, F. C. Grozema, C. F. Guerra, F. M. Bickelhaupt, F. D. Lewis, Y. A. Berlin, M. A. Ratner, L. D. A. Siebbeles, Journal of the American Chemical Society, 127, 148094–14903 (2005).
[99] J. Wang, P. Cieplak, P. A. Kollman, Journal of Computational Chemistry, 21, 1049–1074 (2000).
[100] I. Perez, A. Marchan, D. Svozil, J. Sponer, T. E. Cheatham, C. A. Laughton, M. Orozco, Biophysical Journal, 92, 3817–3829 (2007).
[101] D. van der Spoel, E. Lindahl, B. Hess, G. Groenhof, A. E. Mark, H. J. C. Berendsen, J. Comput. Chem., 26, 1701–1718 (2005).
[102] , W. L. Jorgensen, J. Chandrasekhar, J. D. Madura, R. W. Impey, M. L. Klein, J. Chem. Phys., 79, 926–935 (1983).
[103] X. Lu, M. A. E. Hassan, C. A. Hunter, Journal of Molecular Biology, 273, 681–691 (1997).
[104] C. R. Calladine, H. R. Drew, Understanding DNA; The Molecule & How It Works (Academic Press, London, 1992).
[105] P. W. Anderson, Physical Review, 109, 1492–1505 (1958).
[106] V. M. Kenkre, D. W. Brown, Physical Review B, 31, 2479–2487 (1985).
[107] V. M. Kenkre, D. Schmid, Physical Review B, 31, 2430–2436 (1985).

[108] H. Haken, P. Reineker, Zeitschrift für Physik A Hadrons and Nuclei, 249, 253–268 (1972).
[109] R. Kühne, P. Reineker, Zeitschrift für Physik B Condensed Matter, 22, 201–210 (1975).
[110] S. S. Skourtis, I. A. Balabin, T. Kawatsu, D. N. Beratan, Proceedings of the National Academy of Sciences USA, 102, 3552–3557 (2005).
[111] E. Gudowska-Nowak, Chemical Physics, 212, 115–123 (1996).
[112] I. A. Goychuk, E. G. Petrov, V. May, Journal of Chemical Physics, 103, 4937–4944 (1995).
[113] Y. A. Berlin, F. C. Grozema, L. D. A. Siebbeles, M. A. Ratner, Journal of Physical Chemistry C, 112, 10988–11000 (2008).
[114] M. Galperin, A. Nitzan, M. A. Ratner, Physical Review B, 73, 045314–045326 (2006).
[115] U. Weiss, Quantum Dissipative Systems, Series in Modern Condensed Matter Physics (World Scientific, 1999).
[116] A. Troisi, D. L. Cheung, D. Andrienko, Physical Review Letters, 102, 116602–116605 (2009).
[117] T. F. Soules, C. B. Duke, Physical Review B, 3, 262–274 (1971).
[118] D. Q. Andrews, R. P. Van Duyne, M. A. Ratner, Nano Letters, 8, 1120–1126 (2008).
[119] A. Troisi, M. A. Ratner, M. B. Zimmt, Journal of the American Chemical Society, 126, 2215–2224 (2004).
[120] D. J. Bicout, M. J. Field, Journal of Physical Chemistry, 99, 12661–12669 (1995).
[121] C. Gollub, R. G. S. Avdoshenko, Y. A. Berlin, G. Cuniberti, Israel Journal of Chemistry, 52, 452 (2012).

Biographies

Gianaurelio Cuniberti holds the Chair of Materials Science and Nanotechnology at the Dresden University of Technology and the Max Bergmann Center of Biomaterials Dresden since 2007. He studied Physics at the University of Genoa and at the University of Hamburg and was visiting scientist at MIT and the Max Planck Institute for the Physics of Complex Systems Dresden. From 2003 to 2007 he was at the head of a Volkswagen Foundation Junior Research Group at the University of Regensburg. His activity addresses four main lines: (i) molecular and organic electronics, (ii) bionanotechnology, (iii) nanostructures, (iv) methods development. His research activity is internationally recognized in more than 150 scientific papers to date. He is visiting Distinguished Professor at the Division of IT Convergence Engineering of POSTECH, the Pohang University of Science and Technology and Adjunct Professor for the Department of Chemistry at the University of Alabama.

Rafael Gutierrez graduated in Physics in 1990 and received his PhD at the Dresden University of Technology in 1995. He is currently leader of the Bioelectronics and Neuromorphic Materials Group at the Chair of Materials Science and Nanotechnology at the Dresden University of Technology and the Max Bergmann Center of Biomaterials Dresden. He is author of more than 50 scientific papers and reviews. His current research interests include charge transport in biomolecular systems, spin-dependent transport in helical systems, nanoscale thermoelectrics, and topological insulators.

Interaction of DNA Bases with Gold Substrates

Marta Rosa, Wenming Sun and Rosa Di Felice*

Center S3, CNR Institute of Nanoscience, Via Campi 213/A, 41125 Modena, Italy
**Corresponding author: rosa.difelice@unimore.it*

Received 6 December 2012; Accepted 17 December 2012

Abstract

The interaction of molecules with inorganic substrates is a crucial issue for applications in molecular electronics. It influences important factors such as the immobilization efficiency and the charge injection through the interface. Moreover, mechanical aspects connected to the unfolding of biological molecules are important.

We hereby present recent efforts in our group to tackle these problems, based on density functional theory calculations. In particular, we discuss our results on the adsorption of cytosine on Au(111) and on the interaction of guanine, in its natural and size-expanded forms, with small Au clusters. We find that cytosine binds to the Au(111) surface with a mechanism that involves charge sharing, intermediate between chemisorption and physisorption. The investigation of small complexes between guanine and gold clusters reveals the formation of hydrogen bonds: these configurations with unusual bonds are relevant at the corners of nanoparticles, while they can probably be neglected when DNA binds on flat extended metal surfaces.

Keywords: DNA/Au interfaces, Density Functional Theory, Van der Waals, DNA modifications, electronic hybridization.

Journal of Self-Assembly and Molecular Electronics, Vol. 1, 41–68.

1 Introduction

The electronic structure of DNA molecules started to attract the attention of scientific research groups because of its implications in DNA damage and repair. Prompted by encouraging results from measurements of fast charge transfer in ensembles of DNA molecules in solution, scientists later started to query the possibility to exploit DNA molecules in nanotechnology.

The pioneering experiments on the conductivity of DNA between electrodes were published between 1998 and 2000 [1–4] and showed a variety of electrical behaviors. Since then, much conceptual progress has been made in rationalizing the diverse results [5, 6], by understanding the relevance of surface deposition, electrode contacts, probing single molecules rather than uncontrolled networks and many other factors implied in handling flexible soft biological molecules rather than stiff inorganic (or even organic) materials. For instance, two beautiful experiments were realized in more recent years by avoiding the deposition on a hard substrate and optimizing the electrode-DNA contacts through thiol linkers [7, 8]. Furthermore, new avenues for the exploitation of electrical conductance through DNA have been opened [9].

Our group has intensely worked in the past 10 years on the characterization of the structure and electronic properties of DNA fragments by means of Density Functional Theory (DFT) calculations and Molecular Dynamics (MD) simulations, in close collaboration with experimental groups for interpretation of available data and guidance of new experimental routes (e.g., see our joint exp-theo publications [10–14]). In this short review we focus, instead, on more recent work aimed at addressing the issues of surface deposition and electrode contacts [15, 16]. These issues are closely related to the title of the journal and of the SAME conference that launched the series ("Self-Assembly and Molecular Electronics", Aalborg, Denmark, October 11–12, 2012).

Before doing so, we briefly summarize the basic concepts that emerge from our studies of natural and modified DNA molecules in the gas phase and in solution. We are interested in modified DNAs because chemical and structural alterations may be an effective way to manipulate the electrical behavior of nucleic acids. We argue that natural double-stranded DNA (dsDNA) does not sustain coherent transport through a band-like mechanism, because the electronic structure does not reveal dispersive bands [13, 17]; yet, the electron states are delocalized through the base stack and this superposition may support a form of charge mobility. Continuous wires made of G4 tetrads have a higher density of states close to the fundamental energy gap and a higher amplitude of the occupied states [18, 19], which makes them more promising than dsDNA for exploitation in nanotechnology [10, 20]. Hybridization of each pair in dsDNA with a metal ion is a tool to modulate the fundamental energy gap between electron states [21]. A particularly appealing chemical modification of DNA consists of inserting a benzene ring coplanar to the heterocycles of each base, to form size-expanded xDNA [22]. We have invest-igated the ground-state and excited-state electronic structure of this variant [23, 24] and compared the results to natural dsDNA: we find that, consistently with the intuition of enhanced aromaticity due to the benzene cycle, the π-π coupling between stacked bases increases, with consequences on the ability to conduct charges. Our most recent work is devoted to a viable protocol to take into account the structural flexibility in a solution environment in the computation of the electronic structure of DNAs by high-level DFT methods. We have noted, on the stream of other authors [25–28], that a single frozen structure is not enough to determine the electronic properties of DNA to a quantitative precision that is desired to compare with real experimental data: in fact, different structures may give very different results [23, 24] and the fluctuations should be sorted out for an average picture. Therefore, we are now applying a multi-step procedure that consists of: (1) MD simulation of a DNA (or modified DNA) oligomer in explicit solvent for 100–500 ns; (2) DFT or time-dependent DFT (TDDFT) calculations on selected structures from the MD trajectory. Step 1 is preceded by the DFT-based generation of classical force fields in the case of DNA derivatives for which parameters are missing in the existing public libraries. To go from step 1 to step 2, one would in principle like to sample the trajectory at regular time interval, so that the electronic quantities can be eventually averaged in time. Driven by the need for restricting the number of selected structure to a small value, we opt instead for a different route: we sort the trajectory by a clustering method to extract few representative structures. From these structures, which are not regular snapshots at given times, we are not able to compute a time average of the desired quantities (e.g., optical spectra or transfer integrals); yet, we can understand the structural variability and recover an average picture by exploiting the occurrence of each representative structure along the trajectory. This multi-step protocol has been recently applied to the optical properties of triplex DNA [29] and to the transport properties of a modified duplex DNA [30]. We are continuing to apply and validate the methodology on other appealing modified DNAs.

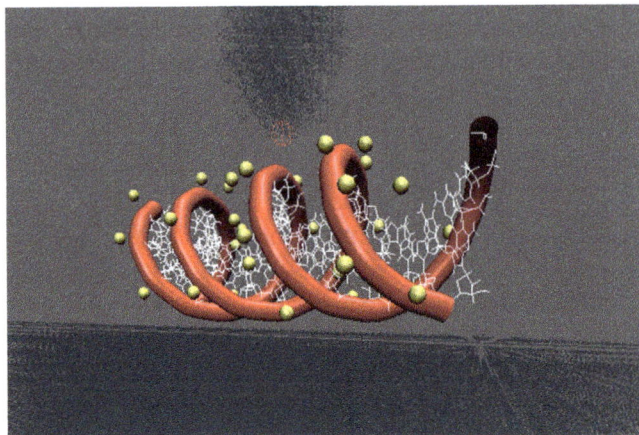

Figure 1 Three-dimensional scheme of a scanning tunneling microscopy experiment on a single DNA molecule. The target molecule lies horizontally on a metal surface: it is still debated whether the double helix remains intact and understanding this issue is a main goal of our research. In other kinds of experiments, thiol linkers are employed to hook DNA to electrodes at the 3' and 5' ends: we want to understand the role of the length and chemical nature of the linkers in the charge transfer between the electrodes and the molecule.

The above background is not intended as a comprehensive survey of the field, but as a brief summary of our knowledge and understanding based on our own research experience. A broader viewpoint on the state-of-the-art theoretical/computational knowledge on the electronic structure of DNA in the perspective of nanoelectronics can be gathered from a good recent review [31]. We now turn to the core topics of this article. One peculiar aspect of nanotechnologies with biological molecules is that they are usually manipulated in setups different from the biological environment. The interaction between biomolecules and inorganic surfaces and nanoparticles is of great importance in natural systems and for the design of bio-nano molecular devices. DNA molecules are deposited onto inorganic substrates for imaging experiments, contacted by metallic tips for spectroscopy and transport experiments [6] (Figure 1). What happens at these interfaces is a major factor to determine the conductance of DNA molecules in a non-native environment and to interpret real experiments. Our early steps towards a thorough investigation of these problems are reported in Sections 2 and 3. In Section 4 we draw our conclusions and consider possible developments in the field. We focus our attention on gold as a substrate because, due to its chemical inertness and biocompatibility, it could be utilized in several bio-electronic and bio-medical applications.

2 The Interaction of DNA Bases with a Flat Gold Surface: The Case of Cytosine on Au(111)

2.1 The Strategy

The theoretical investigation of the fate of a DNA molecule on a hard substrate, such as in the conditions for scanning tunneling microscopy (STM, Figure 1) or atomic force microscopy (AFM) in their various implementations, is not an easy task. One recurring question that still remains unanswered is whether the double helix unfolds. No single computational technique is suitable to accurately deal with all the different length and time scales of the problem.

On the one hand, it is important to take into account the interaction of each DNA base with the surface at the quantum level, because the critical balance of base-base and base-substrate interactions will determine the preference for folded molecule versus a situation in which the nucleotides can interact more strongly with the host substrate: this length scale can be tackled by means of Density Functional Theory to determine geometry variations in the base and in the substrate upon adsorption, the behavior of the electrons with possible charge transfers and the accurate energetics. On the other hand, unfolding is a process that involves large objects and long times, which are not accessible by ab initio electronic structure methods.

To gain fundamental insights into the problem of DNA/surface coupling, we adopted a multi-step approach that was recently proposed for proteins on surfaces [32], based on the combination of DFT calculations, MD simulations and docking simulations. Thanks to the energetic and structural results obtained with DFT calculations of individual bases on a surface of choice, it is possible to develop a classical force field for atomistic MD simulations of an entire DNA oligomer on the surface. While reliable classical force fields are available for nucleic acids and inorganic materials separately, parameters to describe the interaction have not been developed to date. The DFT results are not only instrumental to force field development, but bring insights into the specific system: in particular, they allow us to unravel the mechanisms of adsorption of individual bases, which make a debated issue in the context of self-assembly and molecular electronics with a variety of molecular components. MD simulations of a DNA oligomer on a substrate are feasible for several different sequences. The results of such simulations performed with the new classical force field can in turn promote the setup of large-scale docking simulations [32]. Docking simulations are able to reveal a plethora of viable docking geometries, which can eventually be tested and refined by MD and DFT. This repeating cycle will eventually allow us to unravel the conformation of DNA molecules on an inorganic substrate.

To start this ambitious plan, we have selected the Au(111) surface as the substrate, because it is experimentally interesting and already well characterized for MD simulations. To derive the classical force field for DNA/Au(111) MD simulations, it is necessary to perform DFT calculations of each individual base (cytosine, guanine, thymine and adenine) on Au(111). Here we present our DFT results of the cytosine/Au(111) system as a prototype, to demonstrate the two-fold predictive power of this approach: (1) to gain insights in the electronic structure of the system itself; (2) to develop tools (namely, force field parameters) that enable the simulation of more complex systems. We discuss the microscopic nature of the interaction between cytosine and the Au(111) surface. We particularly focus on quantifying the importance of dispersion interactions in order to develop a reliable multi-step method [32].

We investigated cytosine on Au(111) including the van der Waals (vdW) interaction, which was not done in previous studies of the same system [33]. We could do so by benefiting of a recent implementation of a vdW-corrected functional for DFT calculations, named vdW-DF [34, 35]. Indeed, we reveal significant differences comparing the results obtained with and without vdW interaction, both in total energy values and in equilibrium geometries. Our vdW-DF results are in good agreement with experimental data, while DFT results without vdW terms are affected by larger deviations. These results have implications for monolayer formation and contradict a common opinion that the surface-molecule interaction gives only a minor contribution to the monolayer formation energy. In fact, authors have usually neglected the interaction of each base with the surface when dealing with supramolecular structures [36]. Our work, instead, reveals that, even if vdW terms are essential to attain a correct quantitative and qualitative description, the interaction mechanism goes well beyond pure vdW coupling.

2.2 The Method

We performed gradient corrected DFT calculations of cytosine in the gas phase and adsorbed at the Au(111) with the quantum-espresso package [37]

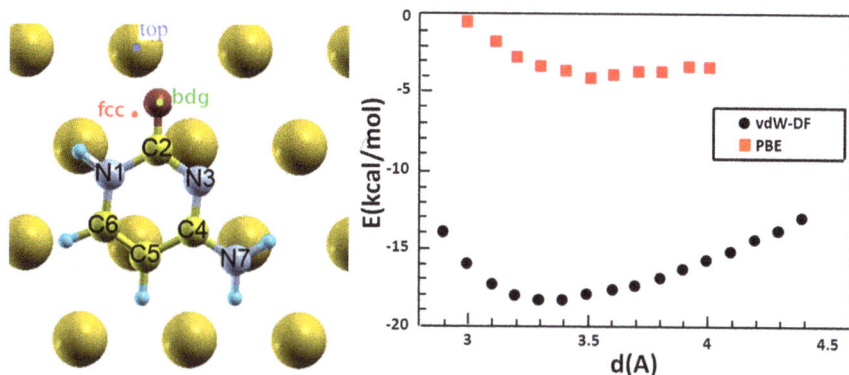

Figure 2 (Left) Top view of a relaxed configuration of cytosine/Au(111). The image defines the atomic labels in the adsorbate and the labels of adsorption sites in the substrate. (Right) Adsorption energy versus cytosine-Au vertical distance (measured between the center of mass of the molecule and the average surface height) for horizontal cytosine on Au(111) at the bdg site. Black and red points denote vdW-DF and DFT [33, 42] results, respectively.

(version 4.3), using the PBE exchange correlation functional [38] and the vdW-DF functional [35, 39]. We chose a plane wave basis set with a cutoff of 25 Ry and we described the electron-ion interaction with ultrasoft pseudo-potentials [40]. The surface was modeled with a slab of four Au layers with a periodically repeated $6 \times 3\sqrt{3}$ surface supercell that contains 36 atoms per layer. The lateral distance between two neighboring cytosine replicas was at least 11 Å and the vacuum thickness in the direction perpendicular to the surface was at least 14 Å, enough neglect the interaction between periodic replicas of the system. The Brillouin zone (BZ) sums were calculated including two Monkhorst–Pack special k points in the irreducible wedge. All the atomic coordinates were relaxed until the forces vanished within 0.05 eV/Å. Preliminary calculations on benchmark systems allowed us to assess the reliability of the method [15].

Atomic charges and the amount of charge transfer were evaluated through Løwdin's population analysis [41].

In the following, we denote with the PBE/vdW-DF (PBE/PBE) label calculations that were done with the PBE functional in the pseudopotential generation and vdW-DF (PBE) functional in the self-consistent runs for the interface. For tests of this ansatz we refer the reader to our original work [15].

Table 1 Adsorption energies, molecule-surface distances and inclinations for cytosine adsorbed on Au(111) [15]. The adsorption site is the site of the Au(111) triangular lattice on top of which the O_2 atom of cytosine resides.

		Adsorption site	Optimized vertical distance	Optimized tilt angle relative to normal	Adsorption energy
			Å	degrees	kcal/mol
Horizontal cytosine/Au(111)	PBE/vdW-DF	bdg	3.2	86	18.5
		fcc	3.2	81	18.5
		top	2.7	76	19.3
	PBE/PBE	bdg	3.3	90	4.5
Vertical cytosine/Au(111)	PBE/vdW-DF	top	2.6		18.6
	PBE/PBE	top	2.3		15.5

2.3 Structure and Energetics

Different adsorption sites and adsorption geometries were tested for the cytosine/Au(111) interface. We performed vdW-DF calculations for geometry optimization starting from cytosine either parallel (horizontal configurations) or perpendicular (vertical configurations) to the substrate. The O atom of cytosine in the initial conditions was placed above a top (one-fold coordination) bridge (abbreviated bdg, two-fold coordination), or fcc (three-fold coordination) site of the underlying triangular lattice (Figure 2).

We started optimizing the atomic coordinates of the horizontal cytosine/Au(111) system both with and without vdW interactions, starting from an arbitrary configuration with the molecule parallel to the surface at 3.4 Å above Au(111). After the relaxation we performed a series of single-point calculations at frozen internal coordinates by varying the molecule-surface vertical distance. The results are visualized in the plot of Figure 2. The adsorption energy values, which are given by the minima of the curves, are sensibly different in vdW-DF and PBE calculations. PBE calculations give a very shallow minimum, while vdW-DF results produce a deeper minimum at the distance of 3.3 Å, which is more compliant with a variety of results on aromatic systems and heterocycles [43, 44]. Our vdW-DF results are in good agreement with experimental data that report adsorption energy gains between 25 and 36 kcal/mol in the high-coverage regime, while PBE calculations deviate from the ones in [42]. The results are summarized in Table 1.

First of all we note that the adsorption energies for cytosine adsorbed in a horizontal mode at different lattice sites differ by less than 1 kcal/mol, namely they can be considered as degenerate. This qualitative feature also results from calculations without the vdW interaction [45]. It also emerges from AFM results on monolayers, both of cytosine and of other DNA bases, which show that the geometry of the monolayer is not influenced by the geometry of the Au(111) lattice [36].

Furthermore, we point out that in PBE results the vertical configuration is definitely preferred with respect to the horizontal one, while with the vdW-DF functional the two configurations have very close adsorption energies and the horizontal one is slightly preferred. This result is in agreement with experimental results that show that cytosine is not able to create a self-assembled monolayer on, e.g., Au(111). Instead, at low coverage cytosine prefers filament structures with a high mobility [36]. In these structures cytosine molecules are always adsorbed horizontally. It is true that the interaction with the other molecules in the filament determine the preferred orientation, but it is also true that, in the presence of a strong preference for a vertical configuration, as for cytosine on Cu(110), monolayers are formed by vertical molecules [36].

The final vdW-DF adsorption energy value of cytosine on Au(111), both in parallel and vertical configurations, gives a remarkable evidence of adsorbate-substrate interaction beyond the pure dispersion regime, despite the fact that dispersion interactions are crucial for a correct description of the system. Another evidence of the interaction between cytosine and Au(111) is the fact that in vdW-DF horizontal configuration results the molecule is tilted respect to the surface and the O atom gets close to the surface. C_2-O_8 bonding is stretched and the cytosine molecule gains 0.2 e$^-$. Thus, our results indicate that electron transfer occurs from the surface to the adsorbate. We believe that the inclusion of dispersion interactions in the theoretical description is crucial to resolve these subtle effects. More insights on the interface coupling are gained by inspecting the electronic structure, which is presented in the next subsection.

2.4 The Electronic Structure and Binding Mechanisms

To understand what kind of interaction is established between the molecule and the surface, we analyzed the density of states (DOS) and the isosurface plots of electron wavefunctions, according to an established protocol [46].

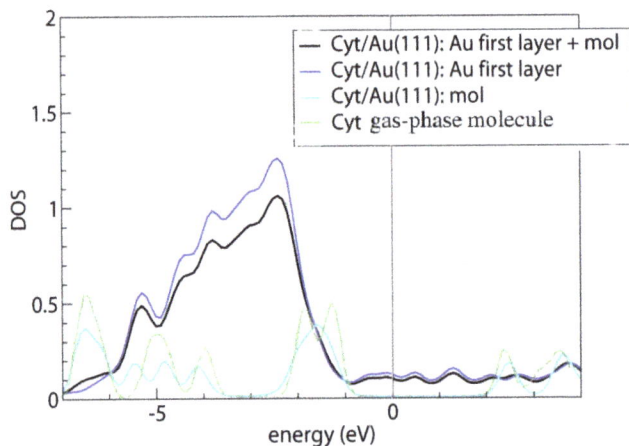

Figure 3 Density of states for cytosine adsorbed on the Au(111) surface, computed by vdW-DF. The Fermi level is set at the origin of the energy scale. The deepest energy level, which is the same for the gas-phase molecule and for the adsorbed system, is used for alignment of the various curves. The green line is the DOS of the gas-phase molecule. The cyan, blue and black lines represent the projections of the total cytosine/Au(111) DOS on cytosine, the top Au layer and the sum of cytosine with the top Au layer, respectively, as indicated in the legend. The projected DOS is computed by projecting the total DOS onto atomic orbitals and then summing over all the projections that constitute the subsystems of interest [15]. Reprinted from [15] with permission; ©2012 American Chemical Society.

Figure 3 illustrates the DOS plots from our vdW-DF calculations of gas-phase and adsorbed cytosine. The green curve is the total DOS of the gas-phase molecule. The other curves are projections of the total DOS of the interface system (the most favorable configuration) o various portions of the compound, as detailed in the figure caption and legend. By comparing the green and cyan lines, we can see only minor perturbations on cytosine in going from the gas phase to the adsorbed phase. In particular, around the upper edge of the Au d bands, namely between 2 and 1 eV below the Fermi level, the DOS shows a redistribution of the cytosine peaks. Since the short distance between the O_2 atom and Au(111), together with the tilted geometry of the molecule, suggest an interaction stronger than pure dispersion, we inspected the energy ranges where both Au and the molecule have a non negligible DOS to search for hybrid orbitals and possibly bonding orbitals.

Figure 4 shows isosurface plots of relevant orbitals of the cytosine/Au(111) interface. The systematic analysis of all the single-particle electron wave functions (we visualized many more orbitals than

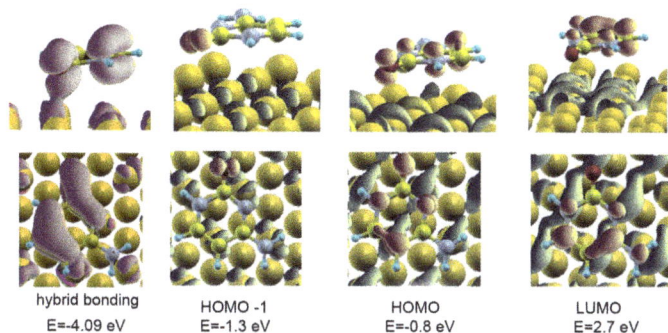

Figure 4 Isosurface plots of representative hybridized orbitals of cytosine adsorbed on Au(111) at the bridge site of the triangular substrate lattice. The leftmost panel shows an example of a bonding orbital between cytosine and the host surface, with a pink isosurface. The other panels show examples of hybrid, but not bonding, orbitals formed with the Au surface by the HOMO-1, HOMO and LUMO: the isosurface portion localized on the surface is blue, the isosurface portion localized on the molecule is red [15]. Reprinted from [15] with permission; ©2012 American Chemical Society.

those reported here) reveals the formation of bonding orbitals, of which the leftmost panel in Figure 4 encloses just an illustrative example.

The Newns–Anderson model for atomic and molecular chemisorption on metal surfaces predicts that the interaction between the localized molecular orbitals and the narrow d band of the metal produces hybrid orbitals of both bonding and antibonding type [47]. This mechanism is found, e.g., in thiols chemisorbed on Au(111) [48] and was accurately discussed in the case of cysteine/Au(111) [46]. The case of the cytosine/Au(111) interface does not comply with the Newn–Anderson picture. We reveal the formation of bonding orbitals but there is not the formation of bonding-antibonding couples from the same molecular orbital and the shift in energy of the bonding orbital respect to the gas-phase molecule is less than 0.5 eV. Only in the energy range of the Au d bands, where molecular orbitals have a component on O_2 or N_3 atoms, the interaction is strong enough to allow charge sharing and the formation of bonding orbitals.

Our findings point out a kind of interaction that is stronger than pure physisorption and that causes the formation of hybrid orbitals between the molecule and the surface. The analysis of vdW-DF results shows that the interaction between cytosine and Au(111) is beyond the pure dispersion regime, even if dispersion interactions are fundamental to bring the molecule close enough to the surface to exploit the short range interaction between O_2,

N_3 and Au. This interaction is then responsible of the tilted geometry of the horizontal molecule and of the formation of bonding orbitals. This evidence, along with results accumulating from recent works [49–52], contradicts the common belief that homo- and hetero-cycles adsorb on metal surfaces by solely dispersion interactions.

2.5 Future Developments

Similar calculations as those presented above have been carried out for all the bases. They allowed us to develop a preliminary DFT based force-field for classical molecular dynamics simulations of DNA on Au(111): this force field is in the final phase of testing before release. The next release of the force-field will include a refinement to take into account the interaction of the sugar-phosphate backbone with the substrate: we expect this correction to be minor, because of weak coupling. In the long term, we plan to proceed along the conceptual strategy outlined in Section 2.1.

3 The Interaction of DNA Bases with Nanostructured Gold: The Case of Guanine at Small Au Clusters

3.1 Motivation and Background

In various investigations of DNA and peptides on Au(111), we recently noticed an unexpected degree of electronic coupling. For example, the dative-bond between a hydroxyl-rich beta-sheet and a gold surface is the essential component of the recognition mechanism [53]. Furthermore, we revealed (section 1.2) electronic hybridization at the cytosine/Au(111) interface [15]. Experimental methods have confirmed that in low coverage DNA strain could wrap the gold nanoparticles [54–56]. It is well known that the adsorption of biomolecules on gold surface/nanoparticles is actually a dynamic process, which depends on the details of adsorbate nature, molecule coverage and environment. DNA bases can adsorb on nanoparticles in various configurations such as parallel and tilted and employing different molecular atoms/groups, depending on the potential energy surface. Here we focus on the formation of unusual H-bonding between guanine (G) and size-expanded guanine (xG) with gold clusters, as a model of the situation at the edges of nanoparticles.

In 2005, Kryachko and Remale investigated the interaction between guanine and small gold clusters by DFT [57]. They determined that the bonding occurs via the N and O atoms in the base and one gold atom. They also showed that this chemical bonding could be reinforced by NH···Au uncon-ventional hydrogen bonds nearby. After that, there were several reports about the unconventional hydrogen bonds in the gold based complexes. Shukla and coworkers investigated the interaction of the guanine base and the Watson–Crick guanine-cytosine base pair with larger gold clusters Au_n ($n = 2, 4, 6, 8, 10,$ and 12): their results confirmed the bonding between N and Au atoms [58]. Other groups investigated the interaction between DNA bases [59] or base pairs with the Au20 cluster [60]. Recently, Cao and coworkers stud-ied the nucleobase-gold complexes with anion photoelectron spectroscopy and DFT calculations and confirmed the existence of NH···Au hydrogen bonds through experimental measurements [61]. In all the past studies, the authors characterized the binding sites and electron affinity/ionization pro-cesses, while mostly omitting information on the detailed topology of the NH···Au hydrogen bonding and N/O-Au bonding. Some questions naturally arise: What is the nature of these interactions? How do the unconventional hydrogen bonds influence the geometrical and electronic properties of these complexes? Is there a way to tune these special interactions? We believe that answering these questions is relevant to explore biomolecule-gold sur-face/nanoparticle interactions, especially at the edge of nanoparticles and at irregular surfaces. Very recently, it was determined that gold clusters (Au_{18} and Au_{27}) tend to pump electrons to the corner and edge sites, making these sites more electronically active than surface sites [62]. Accordingly, the different electronic reactivity of gold atoms at surface, corner or edge sites of nanoparticles may significantly influence the binding orientation of nucleobases on them and consequently the electronic properties of the res-ulting systems. All these phenomena and questions motivated us to get the insights into these special interactions. A guanine molecule in complex with a Au_3/Au_4 cluster are employed to construct atomic models to investigate this problem. We choose guanine among the DNA bases because, due to its low ionization potential (IP) compared with adenine, thymine and cytosine, is expected to play a major role in DNA conductance and lesion/damage processes. Au_3 and Au_4 represent two typical gold clusters, odd and even. It is well known that there is an odd-even effect for the binding strength in gold clusters.

Based on previous work by Kryachko [57], Shukla [58], Cao [61] and others [59, 60], we deduced an initial scheme for the adsorption process of guanine on a gold substrate: on one hand, π-π stacking between molecule and substrate in a horizontal relative orientation would lead to stability by max-imizing orbital mixing; on the other hand, direct N/O-Au bonding between molecule and substrate, possibly reinforced by unconventional NH$\cdots$Au hydrogen bonds, would play a crucial role in an inclined relative orientation for some special conditions. The goal of our work presented in this section is to understand the nature of G@Au$_3$/Au$_4$ coupling and explore ways to tune the competition between horizontal and inclined orientations. The label G@Au$_n$ (xG@Au$_n$) indicates the complex between G (xG) and the Au$_n$ cluster.

In addition to the G base, we also conside xG. In the past ten years, interest in the use of modified analogs of DNA as templates for growing nanoparticle complexes has increased significantly, aiming to reveal whether these new alternative genetic systems could exist for therapeutic and biotechnological applications. Experimental investigations demonstrated the feasibility and stability of DNA double helices that contain size-expanded base pairs [22, 63]. Theoretical studies have shown that the xDNA bases have more electron conjugation than the natural bases in the highest occupied molecular orbital (HOMO) and the lowest unoccupied molecular orbital (LUMO), which in turn affects the value of HOMO-LUMO gap [24, 64–66]. The greater -conjugation may induce strong π-π coupling between stacked bases and bases pairs, for easier charge transfer [23]. Sharma et al. utilized a DFT method to investigate the interaction between x-bases (xA, xC, xT and xG) and the Au$_6$ cluster with the aim of inquiring on the possibility of enhancing conductivity in size-expanded nucleic acids tagged by gold atoms [67]. Bu's group carried out some work on the rational design of hetero-ring-expanded DNA base analogs and determined that the x-bases may be considered as DNA genetic motifs and serve as building blocks in the development of molecular electronic devices [68, 69]. We take steps from these existing data for a deeper investigation of the bonding topology. We wish to understand if and how chemical modifications of the bases affect the DNA@Au binding and in particular the unconventional H-bonding. Progress in this field will offer more information on the potential molecular wire application of natural and modified DNAs.

3.2 The Method

Geometry optimizations have been done with the B3LYP exchange-correlation functional [70] in its restricted and unrestricted forms, employing the LANL2DZ and 6-31+G* basis sets for Au and the natural guanine/x-guanine, respectively. Frequency calculations at the same level were performed as well, to ensure that the equilibrium systems represent true minima on the potential energy surfaces. No symmetry constraint was imposed during the geometry optimizations. The basis set superposition error (BSSE) was corrected by using the counterpoise procedure of Boys and Bernardi [71]. The natural bond orbital (NBO) analysis of charge population and other electronic properties were studied at the B3LYP/LANL2DZU6-311++G** level. All calculations have been performed with the Gaussian03 suite of codes. Topological properties of the electron density at the bond critical points (BCPs) of the NH$\cdots$Au and OH$\cdots$Au hydrogen bonds were characterized using the Atoms in Molecules (AIM) methodology [72] at the B3LYP/LANL2DZU6-311++G** level.

To analyze and visualize the non-covalent interaction in these systems, the non-covalent interaction index (NCI) approach, developed by Yang et al. [73], was adopted. The Multiwfn software [74] was used to calculate relevant quantities of this approach. More details are reported elsewhere [16].

3.3 Structure and Energetics

3.3.1 Natural Guanine

Our initial neutral G@Au$_3$/Au$_4$ structures were taken from the most stable geometries described by Kryachko and Remacle [57]. Our optimized structures of the neutral G@Au$_3$ and G@Au$_4$ complexes at B3LYP/LANL2DZU6-31+G* level are shown in Figure 5. Tests at the MP2/LANL2DZU6-31+G* level assessed the accuracy of the B3LYP/LANL2DZU6-31+G* results.

In Table 2 we present the some key features of neutral G@Au$_3$ and G@Au$_4$ systems computed by us. We find the length of the N$_3$-Au bond to be 2.164 Å, which is only 0.018 Å longer than in previous work. This tiny difference stems from the different basis sets for treating gold atoms and is not significant. The results reported in Table 2 overall demonstrate a good agreement with the previous theoretical data. According to six criteria to justify hydrogen bond formation, vdW cutoff, red shift of infrared intensity (R_{IR}) and downfield shift of nuclear magnetic resonance (σ_{iso}), Kryachko and Remacle found out that the unconventional NH$\cdots$Au bonds obey all the necessary prerequisites of standard H bonds [57]. On the basis of our results in Table 2, we draw the same conclusion.

In order to assess the bonding energy between nitrogen and gold atoms in the G@Au$_3$ neutral complex, we artificially constrained the complex to a different geometry. Specifically, we rotated the Au$_3$ cluster so that the Au$_3$ triangle becomes perpendicular to the guanine plane, without changing the internal geometry of the cluster and the guanine, nor the N$_3$-Au$_1$ distance.

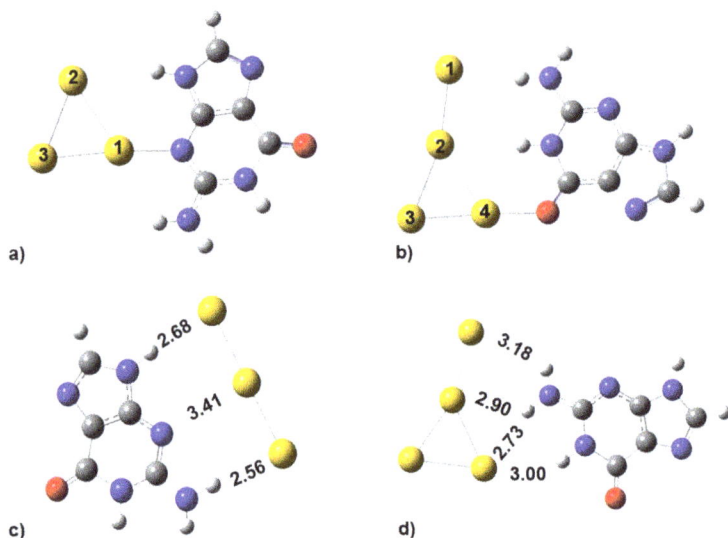

Figure 5 The stable structures for neutral G@Au$_3$ (a), G@Au$_4$ (b), and anionic G@Au$_3^-$ (c), G@Au$_4^-$ (d) complexes.

After this step, the N$_9$H$\cdots$Au$_3$ bond is broken while the N$_3$-Au$_1$ bond is kept. The BSSE corrected binding energy is 21.35 kcal/mol for this geometry, so the N$_9$H$\cdots$Au$_3$ bonding energy is approximately 5.85 kcal/mol. Since the BSSE corrected binding energy for the G-C base pair at the B3LYP/6-311++G**//B3LYP/6-31+G* is 28.4 kcal/mol, the average contribution of each of the three hydrogen bonds in the G-C pair is about 9.5 kcal/mol. Hence, we find that the unconventional hydrogen bond in the neutral G@Au$_3$ complex is weaker than that in a Watson–Crick H-bond in the GC pair.

Due to their large electron affinity, small gold clusters would like to accept an excess electron to be in a more stable anionic state [75]. We want to explore whether the excess electron would increase or decrease the binding strength between gold clusters and the guanine molecule and modify the process of charge transfer in the system.

We optimized the atomic coordinates of the negatively charged G@Au$_3^-$ and G@Au$_4^-$ complexes starting from the equilibrium structures of the corresponding neutral complexes, with one net negative charge added to each system. The optimization process was implemented at the B3LYP/LANL2DZ*U*6-31+G* level as used in the neutral complexes. The BSSE corrected binding energy is 14.24 and 15.37 kcal/mol for the G@Au$_3^-$

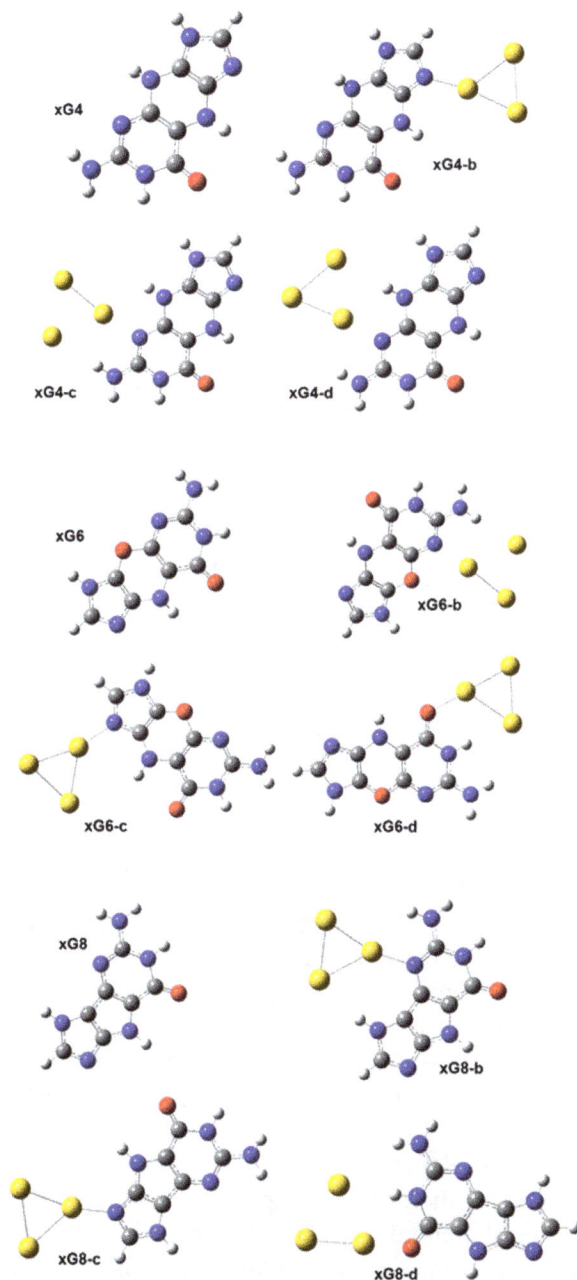

Figure 6 The optimized geometries of the neutral gas-phase xG4, xG6 and xG8 and of the neutral xGn@Au$_3$ complexes ($n = 4$, 6 and 8).

Table 2 Some key features of neutral G@Au$_3$ and G@Au$_4$ complexes. The binding energy (Eb) includes the BSSE correction. It is computed as the total energy of the complex minus the total energies of the isolated guanine and cluster. RIR is the ratio of the IR activities of the N-H stretching mode involved in H-bond in the complex and in isolated guanine. $\Delta\sigma_{iso}$ is the NMR shift (in ppm) taken with respect to the isolated guanine and gold cluster.

Neutral G@Au$_3$			Neutral G@Au$_4$	
	This work	Other theory [57]*		This work
Eb (kcal/mol)	27.2	20.9	Eb	32.01
D(N3-Au1)(Å)	2.164	2.146	R(O6-Au4) (Å)	2.177
ΔR(N9-H9) (Å)	0.010	0.010	ΔR(N2-H2) (Å)	0.009
r(H9…Au) (Å)	2.810	2.841	ΔR(N1-H1) (Å)	0.008
$\angle$N9-H9…Au2 (°)	163.2	161.8	$\Delta\upsilon$(N1-H1) (cm-1)	148.38
$\Delta\upsilon$ (N9-H9) (cm^{-1})	181.8	181	$\Delta\upsilon$(N2-H2) (cm-1)	150.6
R$_{IR}$	6.4	6.0	R$_{IR}$ (N1-H1)	7.6
$\delta\sigma_{iso}$	-1.6	-1.8	R$_{IR}$ (N2-H1)	12.4
			$\Delta\sigma$isoc (N1H)	-2.4
			$\Delta\sigma_{iso}$^{c} (N2H)	-5.4

and G@Au$_4^-$ complex, respectively. These values are about 50% smaller than those found for the neutral complexes. It indicates that the excess electron decreases the binding strength, in agreement with previous indications for different nucleobases. The optimized geometries are presented in Figures 5(c,d), with indication of all the distances between gold atoms and the neighboring hydrogen and nitrogen atoms. In the G@Au$_3^-$ complex the shape of the gold cluster changes from triangular to linear: The angle between the three gold atoms is 178.5°. The N$_3$-Au$_1$ distance increases significantly as compared to neutral G@Au$_3$, by about 1.25 Å. The N$_9$H$\cdots$Au$_2$ distance increases by 0.2 Å and anotherN$_2$H$\cdots$Au$_3$ hydrogen bond forms. To determine the binding energy of the elongated N$_3$-Au$_1$ bond, we adopted the same procedure as described for the neutral complexes, namely rotating the gold rod 90°, while keeping the coordinates of guanine and of the Au$_1$ atom fixed. Thus, the gold rod becomes perpendicular to the guanine plane and the H-bonds are broken. The BSSE corrected binding energy for this structure is 9.56 kcal/mol, which is significantly smaller than its counterpart in neutral complex (21.53 kcal/mol). Therefore, the elongated N_3-Au_1 bond in the $G@Au_3^-$ complex is weaker by ~12 kcal/mol than the corresponding bond in the neutral $G@Au_3$ complex. In the $G@Au_4^-$ complex, the T-shape of the gold cluster does not change significantly as compared to the $G@Au_4$ neutral system. However, the relative guanine/gold orientation changes. We do not find the O_6-Au_1 bond in $G@Au_4^-$, because the gold cluster slides to the N1 and N2 sites of guanine. Hence, there is only hydrogen bonding in the $G@Au_4^-$ cluster, without covalent bonding.

3.3.2 Size Expanded Guanine

We considered two different forms of xG, characterized by different conjugation in the spacer ring [16]: the structures of xG4, xG6 and xG8 are defined in Figure 6. Three initial structures were designed for the xG4@Au_3 complex: the Au_3 cluster has a triangular shape; one Au atom binds to either N7 (Figure 2 xG4-b) or N3 (Figure 2 xG4-c and xG4-d). A NH···Au hydrogen bond was designed in all the initial structures. After optimization, we could not achieve a stable isomer with gold attaching to the oxygen atom of xG4. It should be noted that for the complexes in which the gold-molecule contact is through the N3 site of xG4 (Figure 2, xG4-c,d), the initial geometries are essentially maintained, including the presence of the NH···Au hydrogen bond. However, for the complex in which the gold-molecule contact is through the N7 site of xG4, after optimization the NH···Au hydrogen bond is broken.

Among the three xG4@Au_3 isomers, xG4-b has the lowest total energy, which could be partly explained by the torsion of the xG4 molecule. The xG4 molecule in the gas phase is not exactly planar, with a 166.9° folding angle. The folding angles of xG4 in the three xG4@Au_3 isomers are 168.9, 169.6 and 173.3°, respectively. The BSSE corrected binding energy for the most stable isomer is 27.90 kcal/mol. The energy difference between isomers xG4-c and xG4-d is negligible within the precision of our approach. The fact that the most stable xG4@Au_3 isomer is the one in which the gold-molecule contact is through the N7 site of xG4 and no unconventional hydrogen bond is formed contradicts the expectation that NH···Au hydrogen bonding would reinforce the N/O-Au bond during the combination of cluster and molecule, as we found in the G@Au_3 cluster with natural guanine. We infer that this peculiar behavior of xG4 in complex with a small gold cluster is due to the steric effect caused by the spacer ring.

Compared with xG4, one NH is replaced by oxygen in the spacer ring of xG6. Introducing one oxygen atom in the spacer ring could in principle increase the planarity, but in our results the optimized xG6 is still not planar.

xG8 has, instead, a five-membered spacer ring, which should also facilitate planarity. It should be noted that xG6 has the second lowest HOMO-LUMO gap in the series of size-expanded guanines. Three stable isomers for each of the xG6@Au$_3$ and xG8@Au$_3$ complexes are shown in Figure 6. The isomer xG6-c is the most favorable among the three xG6@Au$_3$ complexes. The BSSE corrected binding energy is 27.04 kcal/mol, very similar to the binding energy of the lowest-energy xG4@Au$_3$ complex. The binding pattern is also similar, with gold-molecule contact occurring at the N7 site of xG6. The isomer xG8-b is the most favorable among the three xG8@Au$_3$ complexes. The BSSE corrected binding energy is 30.42 kcal/mol, higher than that of xG4@Au$_3$ by 2.52 kcal/mol. In this structure, due to the overall good planarity and a reduced steric hindrance, a NH$\cdots$Au can be established to reinforce the N-Au covalent bonding. The xG8@Au$_3$ complex has the largest binding strength of all the xGn@Au$_3$ sampled complexes because of the minor molecular torsion, which in turn enables the formation of NH$\cdots$Au hydrogen bond.

3.4 The Electronic Structure and Binding Mechanisms

3.4.1 Natural Guanine

The results of the NBO analysis and the shape of the frontier molecular orbitals (Figure 7) indicate that each anchoring bond (N$_3$-Au in G@Au$_3$ and O$_6$-Au in G@Au$_4$) involves charge transfer from G to Au. Instead, each unconventional hydrogen bond (one in G@Au$_3$, two in G@Au$_4$) is characterized by charge transfer from gold to G [16].

Charge transfers can be visualized through the electron density difference between the associated state and the isolated individual building blocks. This is shown in the top panel of Figure 8. We find that in the neutral G@Au$_3$ complex the charge variation is mostly due to the covalent N$_3$-Au$_1$ bonding, with minor contributions from the hydrogen bonds.

According to the NBO analysis, the amount of charge transfer from the guanine molecule to the gold cluster is -0.117e and -0.073e in G@Au$_3$ and G@Au$_4$, respectively. Namely, during complex formation electrons are transferred from the molecule to the cluster. The amount of electronic charge transferred to the Au$_3$ cluster is larger than that transferred to the Au$_4$ cluster, due to the different anchoring sites. In the G@Au$_4$ complex, the gold cluster anchors to the oxygen atom of guanine, which has larger electronegativity than the nitrogen anchoring site in the G@Au$_3$ complex. Furthermore, the amount of charge transfer has the same trend as the electron affinity (EA) of

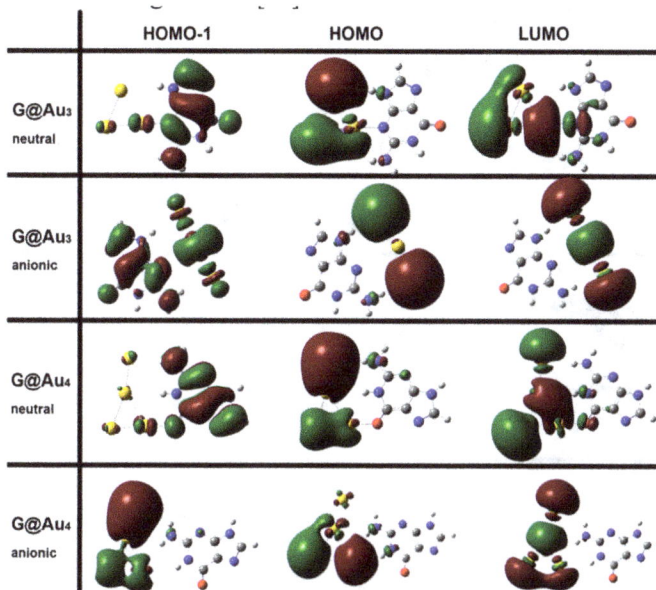

Figure 7 Isosurface plots ($s = 0.02$ a.u.) frontier orbitals for the G@Au$_3$ and G@Au$_4$ complexes in neutral and anionic states.

the Au$_3$ (3.85 eV) and Au$_4$ (2.77 eV) clusters. Our result obtained by the NBO method is consistent with the statement found elsewhere that "gold clusters with an odd number of atoms are better electron acceptors and better electron donors than clusters with and even number of atoms" [60].

The analysis of the charge density and its Laplacian at the bond critical points (BCP) sheds more light into the nature of the unconventional hydrogen bonds in the complexes. Our results indicate that the $(3, -1)$ BCP exists in the NH$\cdots$Au hydrogen bonds and in the N/O-Au bonds. It is easy to comprehend that the nature of the NH$\cdots$Au is a hydrogen bond. What is the nature, however, of the N/O-Au bonds? By analyzing the local energy density $E(r)$, we could determine that the N/O-Au bonds are covalent [16].

It is noteworthy that in the neutral G@Au$_4$ complex, according to the AIM analysis [72], there is no hydrogen bond between N$_1$H and the middle Au$_2$ atom, namely no $(3, -1)$ BCPs. However, according to the same rule we find a hydrogen bond between N$_1$H and the Au$_4$ atom. This is somewhat strange in terms of the atomic distances. In fact, the distance N$_1$H$\cdots$Au$_2$ in the neutral G@Au$_4$ complex is 2.875 Å, larger than the N$_1$H$\cdots$Au$_4$ distance in the same complex but comparable to the N$_9$H$\cdots$Au$_2$ distance in

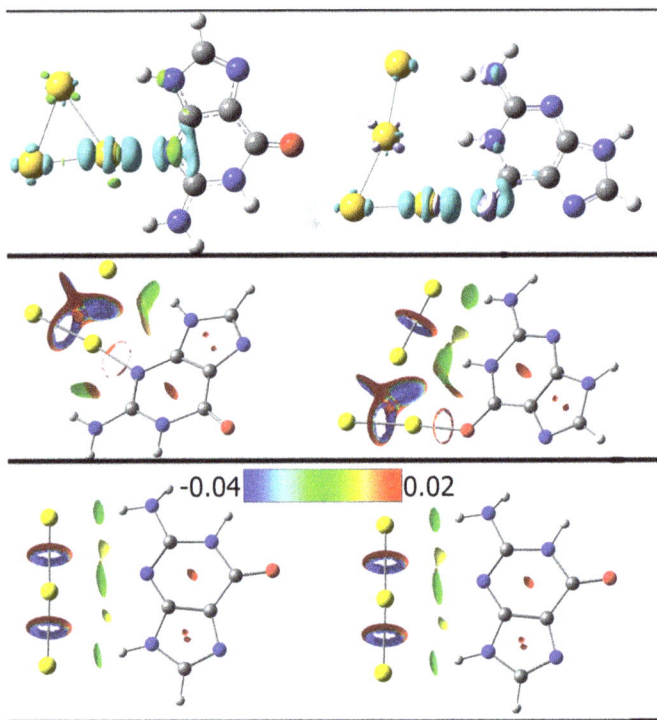

Figure 8 Isosurface plots of the electron density difference (top, $s = 0.005$ a.u.), the reduced density gradient (middle, $s = 0.500$ a.u.) for the neutral G@Au$_3$ and G@Au$_4$ complexes and the anionic G@Au$_3^-$ and G@Au$_4^-$ complexes (bottom, $s = 0.500$ a.u.). The green and cyan isosurfaces in the top plots identify the regions in which the electron density is increased and decreased, respectively, upon complex formation relative to the independent species. The isosurfaces of the reduced density gradient in the bottom plots are colored according to the values of the quantity $\text{sign}(\lambda_2)\rho$ and the RGB scale is indicated.

the G@Au$_3$ complex, and the latter contact was found compatible with H-bonding through the AIM analysis. To further explore the nature of the N$_1$H$\cdots$Au$_2$ contact, we utilized Yang's approach, according to which one can use the sign of λ_2 (the second largest eigenvalue of the electron density Hessian) to distinguish bonded ($\lambda_2 < 0$) from non-bonded ($\lambda_2 > 0$) interactions. The results of this analysis are presented in Figure 8, where the gradient isosurfaces are colored according to the corresponding values of $\text{sign}(\lambda_2)\rho$, which is a good indicator of interaction strength. Large negative values of $\text{sign}(\lambda_2)\rho$ are indicative of attractive interactions (such as dipole-dipole or hydrogen bonding). Large negative values of $\text{sign}(\lambda_2)\rho$ are indicative of non-bonding interactions. Values of $\text{sign}(\lambda_2)\rho$ close to zero indicate weak van der Waals interactions. The isosurface plots shown in the middle of Figure 4 indicate the existence of a H-bond between N_1H and the Au_2 atom in the neutral $G@Au_4$ complex, which instead does not emerge from the BCP analysis. This is a genuine result of our work, which goes beyond previous descriptions. Essentially, we find that Yang's approach performs better than AIM theory in treating this system. The blind point of AIM in the description of the $G@Au_4$ complex may arise from the use of pseudopotentials. The analysis of the frontier molecular orbitals (Figure 7) indicates that both HOMO and LUMO in the anionic complexes are localized on the gold cluster moiety, which could be expected on the basis of the large electron affinity of the gold cluster. The NBO charge population analysis shows that the amount of gold-localized charge in the $G@Au_3^-$ and $G@Au_4^-$ complexes is $-0.927e$ and $-0.936e$, respectively. Also in this case we find a blind point among the BCPs, which can be corrected through the Yang's approach [16]. In fact, the bottom panel in Figure 8 shows the existence of a hydrogen bond between N_2H and the Au_2 atom in the $G@Au_4^-$ anionic complex.

3.4.2 Size Expanded Guanine

The HOMO-LUMO gap for the three $xG4@Au_3$ isomers (b, c and d) is 2.40, 2.22 and 2.45 eV, respectively, which is slightly smaller than that of the $G@Au_3$ complex (2.46 eV).

Comparing the three types of novel size-expanded guanine molecules, we can draw some interesting conclusions. Although the NH site in the spacer ring may supply a new opportunity for forming NH···Au hydrogen bonds in complexes of size-expanded guanine with small gold clusters, this does not occur, because of significant steric effects that hinder further molecule-gold contacts. The introduction of a spacer ring in guanine is an effective tool to decrease the HOMO-LUMO gap and expand the π-conjugation area. These electronic effects were already discussed by other authors [23, 24, 64, 65]. The new characteristic that we find in our work, namely the enhanced steric hindrance, may have other appealing consequences for surface-immobilized DNA-based devices. In fact, our results indicate that the size-expanded guanine molecule favors an adsorption orientation on gold that conduces to π-π stacking, rather than a configuration that maximizes the number of point contacts. The size-expanded design would make the adsorption or assembly process between DNA bases and gold surface/nanoparticle more orderly. As a consequence of all these evidences, we suggest that the new designed size-expanded guanine molecules should have an optimal performance in

DNA-based devices because of (1) the good inherent charge transfer capabilities imparted by the low HOMO-LUMO gap and large -conjugation area [23] and (2) the highly uniform adsorption orientation caused by the enhanced steric hindrance [16]. The latter issue deserves further attention and is the object of ongoing studies. It means that not only the inherent molecular properties are important to develop molecular devices, but also the interaction of the molecules with the inorganic components. Both aspects can be exploited to improve the current state-of-the-art.

3.5 Perspectives

The natural continuation of this work is the investigation of more and more bonding schemes between guanine bases and gold substrates. In particular, we are now focusing on thiol bonds realized by attaching different thiol linkers to a guanine base that is thought as a possible termination of a DNA molecule. Our goal is to identify effects of the chemistry and length of the linker in charge transfer from the substrate (electrode) to the DNA chain.

4 Summary and Outlook

In this article we surveyed our recent work devoted to interfaces between DNA and metal substrates. The substrates that we have in mind can be perfect or defected two-dimensional surfaces, as well as finite-size nanoparticles.

We have revealed important effects of electronic hybridization between nucleobases and Au, similar to the adsorption mechanisms of other homo- and hetero-cycles on Au. Furthermore, we have pointed out that also hydrogen bonding between adsorbate and substrate can be relevant at edges or defects. We have also found that aromatic size-expansion improves not only the electronic structure of the guanine base in view of charge transfer, but also the surface adhesion, which is an additional motivation for its appeal in nanotechnology, besides its obvious impact in biochemistry and molecular biology.

Although we investigated the adsorption of individual bases and not entire DNA oligomers, these results pose the foundations to tackle the more complex interfaces of technological interest.

Before reaching the capability of investigating entire DNA molecules on a surface or a nanoparticle, our short-term interest in the evolutionary line of this work concentrates on developing an AMBER-like force field to enable classical MD simulations of DNA on Au(111) and on probing surface-molecule charge transfer at guanine/Au(111) interfaces realized with different thiol bonding motifs.

Acknowledgements

This work was funded by the European Commission through project "DNA-Nanodevices" (Contract #FP6-029192), by the ESF through the COST Action MP0802, by the Italian Institute of Technology through project MOPROS-URF and the Computational Platform, by Fondazione Cassa di Risparmio di Modena through Progetto Internazionalizzazione 2011. The ISCRA staff at CINECA (Bologna, Italy) is acknowledged for computational facilities and technical support.

References

[1] E. Braun, et al., Nature, 391(6669),775–778 (1998).
[2] P. J. de Pablo, et al., Physical Review Letters, 85(23), 4992–4995 (2000).
[3] H. W. Fink, C. Schonenberger, Nature, 398(6726), 407–410 (1999).
[4] D. Porath, et al., Nature, 403(6770), 635–638 (2000).
[5] R. G. Endres, D. L. Cox, R. R. P. Singh, Reviews of Modern Physics, 76(1), 195–214 (2004).
[6] D. Porath, G. Cuniberti, R. Di Felice, Long-Range Charge Transfer in DNA Ii, 237, 183–227 (2004).
[7] H. Cohen, et al., Proceedings of the National Academy of Sciences of the United States of America, 102(33), 11589–11593 (2005).
[8] B. Xu, et al., Nano Letters, 4(6), 1105–1108 (2004).
[9] M. Zwolak, M. Di Ventra, Reviews of Modern Physics, 80(1), 141–165 (2008).
[10] H. Cohen, et al., Nano Letters, 7(4), 981–986 (2007).
[11] D. A. Ryndyk, et al., Acs Nano, 3(7), 1651–1656 (2009).
[12] E. Shapir, et al., Advanced Materials, 23(37), 4290–4294 (2011).
[13] E. Shapir, et al., Nature Materials, 7(1), 68–74 (2008).
[14] E. Shapir, et al., Journal of Physical Chemistry C, 114(50), 22079–22084 (2010).
[15] M. Rosa, S. Corni, R. Di Felice, Journal of Physical Chemistry C, 116(40), 21366–21373 (2012).
[16] W. Sun, R. Di Felice, Journal of Physical Chemistry C, 116(47), 24954–24961 (2012).
[17] R. Di Felice, et al., Physical Review B, 65(4) (2002).
[18] A. Calzolari, et al., Applied Physics Letters, 80(18), 3331–3333 (2002).
[19] R. Di Felice, et al., Journal of Physical Chemistry B, 109(47), 22301–22307 (2005).
[20] P. B. Woiczikowski, et al., Journal of Chemical Physics, 133(3), Art No. 035103 (2010).
[21] G. Brancolini, R. Di Felice, Journal of Physical Chemistry B, 112(45), 14281–14290 (2008).
[22] A. T. Krueger, et al., Accounts of Chemical Research, 40(2), 141–150 (2007).
[23] A. Migliore, et al., Journal of Physical Chemistry B, 113(28), 9402–9415 (2009).

[24] D. Varsano, A. Garbesi, R. Di Felice, Journal of Physical Chemistry B, 111(50), 14012–14021 (2007).

[25] A. Troisi, G. Orlandi, Journal of Physical Chemistry B, 106(8), 2093–2101 (2002).

[26] R. Gutierrez, et al., New Journal of Physics, 12, Art No. 208102 (2010).

[27] P. B. Woiczikowski, et al., Journal of Chemical Physics, 130(21), Art. No. 215104 (2009).

[28] R. Gutierrez, et al., Physical Review Letters, 102(20), Art. No. 208102 (2009).

[29] T. Ghane, et al., Journal of Physical Chemistry B, 116(35), 10693–10702 (2012).

[30] M. H. Lee, et al., Journal of Physical Chemistry B, 116(36), 10977–10985 (2012).

[31] S. S. Mallajosyula, S. K. Pati, Journal of Physical Chemistry Letters, 1(12), 1881–1894 (2010).

[32] R. Di Felice, S. Corni, Journal of Physical Chemistry Letters, 2(13), 1510–1519 (2011).

[33] S. Piana, A. Bilic, Journal of Physical Chemistry B, 110(46), 23467–23471 (2006).

[34] D. C. Langreth, et al., International Journal of Quantum Chemistry, 101(5), 599–610 (2005).

[35] T. Thonhauser, et al., Physical Review B, 76(12), Art. No. 125112 (2007).

[36] R. E. A. Kelly, et al., Journal of Chemical Physics, 129(18), Art. No. 184707 (2008).

[37] P. Giannozzi, et al., Journal of Physics-Condensed Matter, 21(39), Art. No. 395502 (2009).

[38] J. P. Perdew, K. Burke, M. Ernzerhof, Physical Review Letters, 77(18), 3865–3868 (1996).

[39] M. Dion, et al., Physical Review Letters, 92(24), Art. No. 246401 (2004).

[40] D. Vanderbilt, Physical Review B, 41(11), 7892–7895 (1990).

[41] P. O. Lowdin, Journal of Chemical Physics, 18(3), 365–375 (1950).

[42] M. Ostblom, et al., Journal of Physical Chemistry B, 109(31), 15150–15160 (2005).

[43] S. K. M. Henze, et al., Surface Science, 601(6), 1566–1573 (2007).

[44] F. S. Tautz, Progress in Surface Science, 82(9–12), 479–520 (2007).

[45] S. Rapino, F. Zerbetto, Langmuir, 21(6), 2512–2518 (2005).

[46] R. Di Felice, A. Selloni, E. Molinari, Journal of Physical Chemistry B, 107(5), 1151–1156 (2003).

[47] B. Hammer, J. K. Norskov, Chemisorption and Reactivity on Supported Clusters and Thin Films, 331, 285–351 (1997).

[48] M. C. Vargas, et al., Journal of Physical Chemistry B, 105(39), 9509–9513 (2001).

[49] A. Ferretti, et al., Physical Review Letters, 99(4), Art. No. 046802 (2007).

[50] F. Iori, S. Corni, R. Di Felice, Journal of Physical Chemistry C, 112(35), 13540–13545 (2008).

[51] N. Lorente, et al., Physical Review B, 68(15), Art. No. 155401 (2003).

[52] K. Toyoda, et al., Journal of Chemical Physics, 132(13), Art. No. 134703 (2010).

[53] A. Calzolari, et al., Journal of the American Chemical Society, 132(13), 4790–4795 (2010).

[54] I. Lynch, A. Salvati, K. A. Dawson, Nature Nanotechnology, 4(9), 546–547 (2009).

[55] J. J. Storhofff, et al., Langmuir, 18(17), 6666–6670 (2002).

[56] C. Tamerler, M. Sarikaya, Philosophical Transactions of the Royal Society a-Mathematical Physical and Engineering Sciences, 367(1894), 1705–1726 (2009).

[57] E. S. Kryachko, F. Remacle, Nano Letters, 5(4), 735–739 (2005).

[58] M. K. Shukla, et al., Journal of Physical Chemistry C, 113(10), 3960–3966 (2009).

[59] A. Kumar, P. C. Mishra, S. Suhai, Journal of Physical Chemistry A, 110(24), 7719–7727 (2006).
[60] A. Martinez, Journal of Physical Chemistry C, 114(49), 21240–21246 (2010).
[61] G. J. Cao, et al., Journal of Chemical Physics, 136(1), Art. No. 014305 (2012).
[62] A. Moghaddasi, M. Zahedi, P. Watson, Journal of Physical Chemistry C, 116(8), 5014–5018 (2012).
[63] H. B. Liu, et al., Science, 302(5646), 868–871 (2003).
[64] M. Fuentes–Cabrera, et al., Journal of Physical Chemistry B, 110(12), 6379–6384 (2006).
[65] M. Fuentes–Cabrera, B. G. Sumpter, J. C. Wells, Journal of Physical Chemistry B, 109(44), 21135–21139 (2005).
[66] A. M. Leconte, F. E. Romesberg, Nature, 444(7119), 553 (2006).
[67] P. Sharma, et al., Journal of Chemical Theory and Computation, 3(6), 2301–2311 (2007).
[68] L. Han, et al., Journal of Physical Chemistry B, 113(13), 4407–4412 (2009).
[69] J. M. Zhang, R. I. Cukier, Y. X. Bu, Journal of Physical Chemistry B, 111(28), 8335–8341 (2007).
[70] C. T. Lee, W. T. Yang, R. G. Parr, Physical Review B, 37(2), 785–789 (1988).
[71] S. F. Boys, F. Bernardi, Molecular Physics, 19(4), 553 (1970).
[72] R. F. W. Bader, Chemical Reviews, 91(5), 893–928 (1991).
[73] E. R. Johnson, et al., Journal of the American Chemical Society, 132(18), 6498–6506 (2010).
[74] T. Lu, F. W. Chen, Journal of Computational Chemistry, 33(5), 580–592 (2012).
[75] A. Martinez, Journal of Physical Chemistry A, 113(6), 1134–1140 (2009).

Biographies

Marta Rosa studied Physics in Bologna, Italy, where she received her Laurea in 2010. She is currently a PhD student in Physics in Modena, Italy. Her PhD research program focuses on understanding the mechanisms of formation of complexes between DNA molecules and extended Au surfaces with a multi-step computational approach that ranges from ab initio to empirical docking.

Wenming Sun was born in Feicheng, China, January 1985. He received his B.Sc degree in Chemistry from Shandong University, China. He obtained his Ph.D. in Theoretical and Computational Chemistry, under supervision of Prof. Yuxiang Bu, from Shandong University. He joined Rosa Di Felice's group, as a post-doc at the Center S3, CNR Institute of Nanoscience (Modena, Italy) in November 2011. His research activity is mostly in the field of charge transfer at biomolecule-inorganic interfaces.

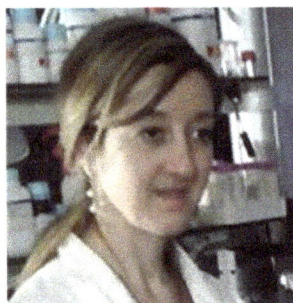

Rosa Di Felice received her Ph.D. in Physics in 1996 at the University of Rome "Tor Vergata", Italy. She is a member of the research staff in the Italian National Research Council (CNR, previously INFM) in Modena, Italy, since 2001. She is an author of about 100 peer reviewed journal papers. Her current research interests focus on the theoretical/computational investigation of the electronic structure of nucleic acids and surfaces and of their complexes with inorganic materials.

Synthesis and Properties of Conjugates between Silver Nanoparticles and DNA-PNA Hybrids

Gennady Eidelshtein[1], Shay Halamish[1], Irit Lubitz[1],
Marcello Anzola[2], Clelia Gannini[2] and Alexander Kotlyar[1,*]

[1]*Department of Biochemistry and Molecular Biology, George S. Wise Faculty of Life Sciences and The Center of Nanoscience and Nanotechnology, Tel Aviv University, Ramat Aviv 69978, Israel*
[2]*Dipartimento di Chimica, Università degli Studi di Milano, via Venezian 21, 20133 Milano, Italy*
Corresponding author: e-mail: s2shak@post.tau.ac.il

Received 5 October 2012; Accepted 19 November 2012

Abstract

We describe the preparation and properties of a stable conjugate between two 15 nm silver nanoparticles (AgNPs) and a DNA-PNA hybrid composed of 10 guanine-cytosine base pairs. We show that the conjugate is spontaneously formed during incubation of a DNA-PNA hybrid, containing phosphorothioate residues at both ends of the DNA strand with AgNPs. The conjugate molecules were separated from individual AgNPs and multiparticle structures by gel electrophoresis. We demonstrate that the absorption spectrum of the conjugate is broader than that of AgNPs, due to the interparticle plasmon coupling.

Keywords: Silver nanoparticles, PNA-DNA hybrid, nanomaterials, TEM.

Journal of Self-Assembly and Molecular Electronics, Vol. 1, 69–84.

1 Introduction

DNA driven self-assembly of nanoparticles has proved to be useful for the synthesis of novel functional nanomaterials [1]. Relatively simple nanostructures, composed of a small number (2 to 10) of nanoparticles connected by DNA [2–7] as well as more complex two- and three-dimensional ones [8, 9] can be fabricated. In addition to unique optical properties, noble metal nanoparticles-DNA conjugates may exhibit electrically conductive or semiconductive behavior and thus, serve as elements in nanoelectronics devices and circuits. Electronic transport measurements on double stranded (ds) DNA molecules have so far yielded very controversial results [10, 11]. Main challenges of direct conductivity measurements are associated with establishing direct physical contact between the DNA and metal electrodes and preserving the native ds conformation of the nucleic acid polymer during conductive measurement under ambient conditions. The environmental factors (e.g. pH, temperature, ionic strength of the medium) greatly affect the stability of the double stranded helix. At low ionic strength the negatively charged strands have a strong tendency to separate from each other. The preparation of samples for direct electrical measurements commonly includes deposition of DNA on electrodes followed by rinsing the surface with distilled water and drying. This treatment can thus lead to the strand separation and, as a result, to a dramatic reduction of the DNA conductivity.

In contrast to dsDNA, stability of PNA-DNA hybrids is independent of ionic strength. PNA (Peptide Nucleic Acid) is a synthetic polymer that composed of nucleic bases connected by peptide bonds [12]. The PNA-DNA hybrid adopts a ds-helical conformation that is very similar with respect to π-π interactions between nucleic bases to a native DNA [13]. PNA is uncharged at neutral pH unlike DNA, and no repulsion occurs between the complementary strands in dsPNA-DNA hybrids. The absence of interstrand repulsion governs high stability of the hybrids under a wide range of experimental conditions and makes the molecule potentially useful for nanoelectronics.

Here we report synthesis of a PNA-DNA hybrid composed of a PNA strand, $(pG)_{10}$ and a complementary DNA strand, $(dC)_{10}$ as well as preparation of conjugates between the hybrid and 15 nm AgNPs. Optical properties and molecular morphology of individual conjugates were measured by absorption spectroscopy and transmission electron microscopy (TEM).

2 Experimental Procedures

Unless otherwise stated, reagents were obtained from Sigma-Aldrich (USA).

2.1 DNA Oligonucleotides

Cytosine-rich oligonucleotides: $[5'-(da)_{10}-(dC)_{10}-(da)_{10}]$, composed of 10 deoxycytidine fragment $(dC)_{10}$ flanked by two runs of phosphorothioated adenosines, $(da)_{10}$ on either side of the strand and $(dC)_{10}$, composed of 10 deoxycytidines, were purchased from Alpha DNA (Canada). These C-oligonucleotides were purified on a C8 4.6 × 250 mm (Supelco Inc.) reverse-phase HPLC column. Elution was performed with a linear Methanol gradient from 0 to 50% in 30 mM K-Pi, pH 7.5 for 85 min at a flow rate of 0.7 mL/min at ambient temperature. Chromatography was performed on Agilent 1100 HPLC system. The oligonucleotides eluted from the column were desalted using a pre-packed Sephadex G-25 DNA-Grade column equilibrated with 2 mM Tris-Acetate, pH 7.5. Quantification of oligonucleotides was done by UV spectrophotometry using extinction coefficients of: 7.4, 15.4 and 11.4 mM^{-1} cm^{-1} at 260 nm for C-, A- and G-bases respectively [14]. Concentration of the $5'-(da)_{10}-(dC)_{10}-(da)_{10}$ was quantified, using an extinction coefficient of 382 $mM^{-1}cm^{-1}$ at 260 nm.

2.2 PNA Oligonucleotide

A PNA strand, containing 10 G-bases, a Fluorescein moiety at N-terminus and a NH_2 group at C-terminus of the sequence, Flu-$(pG)_{10}$, was synthesized on MBHA resin using a 20 μmol scale Boc protocol essentially as described [15]. The synthesis was performed manually in 8 mL reaction vessel, filled with 100 mg of polystyrene resin beads. The synthetic procedure included 10 Boc deprotection-coupling-washing cycles. The synthesized G-decamer was conjugated with N-Fmoc-6-aminohexanoic (Fmoc-Ahx) pre-activated in a similar manner, and the Fluorescein isothiocyanate was subsequently attached to the N-terminus of Ahx. The kinetics of the PNA-oligomer extension and the conjugation reaction were controlled by ESI and MALDI. The product of the synthesis was purified on a 9.4 × 250 mm ZORBAX 300SB-C18 reverse-phase HPLC column (Agilent Technologies, USA). The elution was in 0.1% TFA at a flow rate of 3 mL/min with a linear acetonitrile gradient from 5 to 40%. The oligonucleotide was dried by liophylization. The second purification step included ion-exchange HPLC at alkaline pH. The solubility of Flu-$(pG)_{10}$ in aqueous medium at neutral pH is very low. At alkaline pH the oligonucleotide however becomes highly soluble, as a result of deprotonation of G-bases. Flu-$(pG)_{10}$ was dissolved in 0.5 M LiOH and chromatographed on an anion-exchange HiTrap QHP, 5×1 mL FPLC column (Amersham-Biosciences, USA) in 0.1 M NaOH containing 10% acetonitrile. The elution was with linear NaCl gradient from 0.5 to 1 M at a flow rate of 0.7 mL/min. The major peak fraction was collected, loaded onto a Sephadex NAP-25 DNA-Grade column, 15×50 mm (GE Healthcare, USA), equilibrated with 2 mM Tris-Acetate, pH 7.5 and eluted with the same buffer. The PNA solution was placed into plastic (1.5 mL capacity) tubes and stored at 4°C.

2.3 PNA-DNA Hybrid

To prepare a ds-Flu-labeled PNA-DNA hybrid, Flu-$(pG)_{10}$-$[(da)_{10}$-$(dC)_{10}$-$(da)_{10}]$, HPLC purified Flu-$(pG)_{10}$ and $(da)_{10}$-$(dC)_{10}$-$(da)_{10}$ were mixed at a 2:1 molar ratio in 2 mM Tris-Acetate, pH 7.5. The mixture was heated to 80°C for 15 min and left for 2 to 3 hours at ambient temperature. NaCl was then added to a final concentration of 100 mM. The clear solution turned turbid after 30 min due to precipitation of Flu-$(pG)_{10}$; the hybrid does not precipitate under these conditions. The sample was then centrifuged in a 5424 Eppendorf bench-top centrifuge at 14,000 rpm for 5 min. The pellet containing Flu-$(pG)_{10}$ was discarded and the supernatant, containing the ds-PNA-DNA was transferred to an Eppendorf tube and stored at 4°C. Concentration of the hybrid was quantified, using an extinction coefficient of 497 mM^{-1} cm^{-1} at 260 nm.

2.4 15 nm AgNPs

180 mL of cooled DDW/filtered water were added into a 0.5 L glass beaker placed in an ice-water bath. 0.45 mL of 0.1 M AgNO$_3$, 0.90 mL of 50 mM sodium citrate and 0.75 mL of 0.6 M NaBH$_4$ were consequently added into the beaker under vigorous stirring. The yellow solution was stored at 4°C for 12–16 h. 0.72 mL of 2.5 M LiCl were then added under constant stirring at ambient temperature. The solution was transferred into 15 mL capacity DuPont Pyrex tubes and centrifuged at 14,000 rpm for 1.5 h at 20°C in a Sorval SS-34 rotor. A fluffy pellet was collected. Concentration of AgNPs was estimated spectroscopically using extinction coefficient of 2×10^9 at 400 nm.

Coating of AgNPs with $(dA)_{10}$, an oligonucleotide composed of 10 deoxyadenosines, was conducted as follows: 20 μM $(dA)_{10}$ was added to

4 mL of AgNPs (OD~ 90 at 400 nm). NaCl was then added to a final concentration of 25 mM and the solution was left at ambient temperature for 1 h. The concentration of the salt was increased up to 50 mM. One hour later the concentration of NaCl was adjusted to 100 mM, the solution was incubated at ambient temperature for another 80 min and loaded into a Sepharose 6B-CL column (1.6 × 35 cm). Elution was with 10 mM Na-Pi, pH 7.4 at ambient temperature. The yellow eluate was collected into 1.5 mL Eppendorf tubes and centrifuged at 13,000 rpm for 40 min at RT on bench-top centrifuge 5424 (Eppendorf, Germany). The fluffy pellet was collected and stored in dark at ambient temperature. The resulting nanoparticles were screened for their size and uniformity by TEM, revealing an average diameter of 15 ± 3 nm. The visible spectra showed a characteristic absorption peak at 400 nm. Concentration of the particles was calculated using an extinction coefficient (ε) of 2×10^9 cm^{-1} at 400 nm [16].

2.5 AgNP-DNA-PNA Conjugates

AgNPs were mixed with Flu-(pG)$_{10}$-[(da)$_{10}$-(dC)$_{10}$-(da)$_{10}$] at different molar ratios and incubated for 16 h in 5 mM K-Pi, pH 7.5, containing 100 mM NaCl at ambient temperature. The incubation mixture was electrophoresed on a 1.5% agarose gel. The gel areas were cut out from the gel with a razor blade and the conjugates were electroeluted into dialysis bags. Electroeluted samples were centrifuged at 13,000 rpm for 40 min on a 5424 Eppendorf bench-top centrifuge. The pellets were suspended in a small (20–100 μL) volume of TEA buffer.

2.6 Gel Electrophoresis

Samples were loaded into a 7 × 7 cm 1.5% agarose gel, and electrophoresed at 4°C and 130 V for 45 min, using TAE as a running buffer. The gel was stained with ethidium bromide (5 μg/mL) for 20 min. The DNA was visualized with a Bio Imaging System 202D at 302 nm.

2.7 Absorption Spectroscopy

Absorption spectra were acquired with a UV/VIS Evolution-60 spectrophotometer from Thermo Scientific (USA). Measurements were conducted at 25°C in a wavelength range from 350 to 800 nm.

2.8 HPLC

Chromatography assays were done on Agilent 1100 HPLC system, (Hewlett Packard, USA), including a photodiode array detector unit with quartz flow cells with 1 cm optical path.

2.9 TEM Spectroscopy

TEM images were acquired by using carbon-coated grids (400 mesh). 2.5 μL of a sample in 40 mM Tris-Acetate, pH 7.8, were dropped onto a grid surface. After incubation for 5 min at ambient temperature, the excess solution was removed by blotting with a filter paper. TEM imaging was performed on a TEM (JEM model 1200 EX instrument) operated at an accelerating voltage of 120 kV.

3 Results

3.1 Preparation and Characterization of a DNA-PNA Hybrid

Synthesis of the hybrid includes (see Figure 1): (1) Heating a mixture of HPLC purified PNA and DNA strands (PNA to DNA molar ratio is 2) to 80°C and slow cooling the sample down to ambient temperature; (2) addition of 100 mM NaCl, incubation of the mixture at room temperature for an hour and separation of the hybrid from the excess of Flu-$(pG)_{10}$ by centrifugation.

We have shown that the hybrid elutes as a single peak from a size-exclusion HPLC column (Figure 2A) and is characterized by absorption maxima at 258 and 495 nm (see inset in Figure 2A). The absorption at 490 and 260 nm reflects the relative amounts of fluorescein and nucleic bases in the hybrid respectively. The spectrum corresponds nicely with the molecule composed of 10 G-, 10 C-bases and one fluorescein residue. The hybrid migrates through the agarose gel as a single narrow band (Figure 2B). The results presented in Figure 2 thus show that the hybrid is stable and does not dissociate into single strands during HPLC and electrophoresis.

3.2 Synthesis of Oligonucleotide-Coated Silver Nanoparticles

The citrate-protected 15 nm AgNPs are unstable and precipitate out of the solution at salt concentrations higher than 30 mM. In order to increase their stability we have coated them with an oligonucleotide composed of 10 deoxy-adenines, $(dA)_{10}$. Incubation of AgNPs with large (300–500 fold) molar excess of each of the following sequences: $(dA)_{10}$, $(dC)_{10}$ or $(dG)_{10}$ greatly

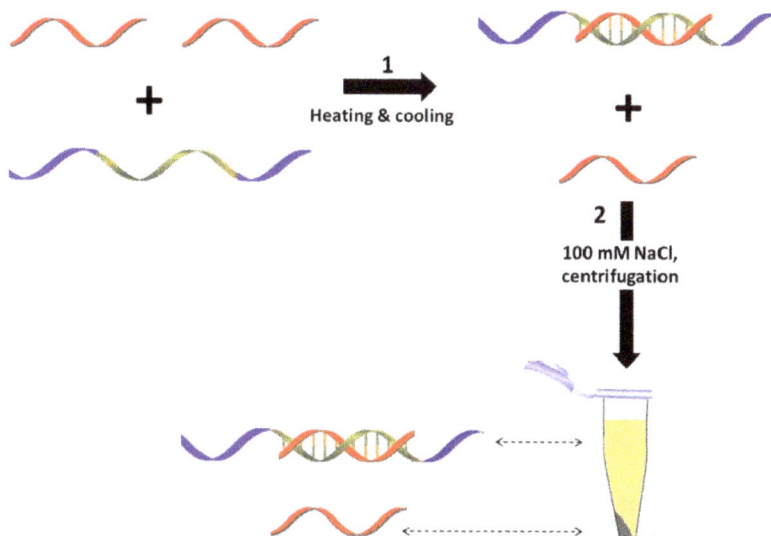

Figure 1 Schematic representation of Flu-$(pG)_{10}$-$[(da)_{10}$-$(dC)_{10}$-$(da)_{10}]$ synthesis. Flu-$(pG)_{10}$ is depicted in red, $[(da)_{10}$-$(dC)_{10}$-$(da)_{10}]$ in green and blue; blue fragments correspond to $(da)_{10}$.

increased the resistance of the particles to salts. The above oligonucleotide-coated particles did not precipitate at 100 mM NaCl in contrast to citrate-protected ones that aggregate at salt concentrations exceeding 30 mM. In contrast, $(dT)_{10}$ was incapable of stabilizing the particles.

We have also demonstrated that incubation with ATP did not affect the resistance of citrate-protected AgNPs to salts. The particles coated with $(dA)_3$, $(dC)_3$ or $(dG)_3$ did not aggregate at 50 mM NaCl, but precipitated at higher salt concentrations (data not presented). These results clearly show that stability of nanoparticles depends on the length of oligonucleotide used for coating.

The $(dA)_{10}$-coated particles can be chromatographed and electrophoresed in contrast to citrate-protected ones that do not enter columns and gels. This property enables one to use size-exclusion chromatography (see Section 2) for purification of the particles from high molecular weight aggregates that are always present in the suspension. The size-exclusion chromatography yielded monodisperse spherical nanoparticles with narrow size distribution as revealed by TEM analysis (see Figure 4A). As seen in the TEM image the average size of the particles is equal to 15 ± 3 nm.

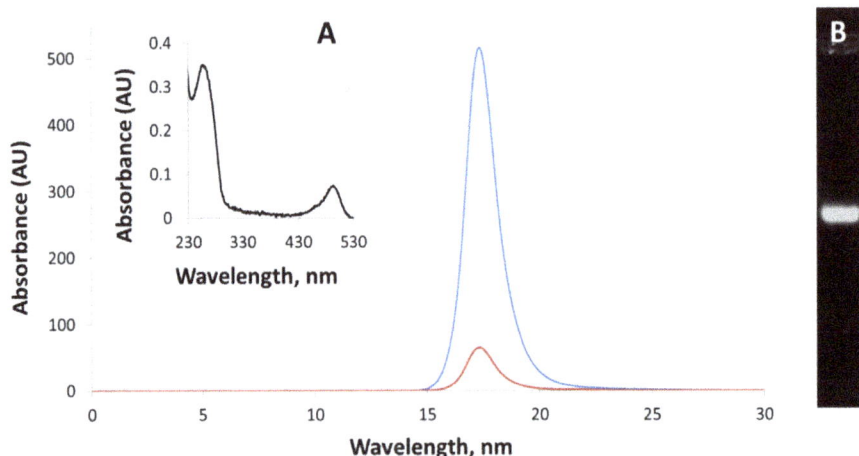

Figure 2 Size-exclusion HPLC (A) and electrophoresis (B) of Flu-$(pG)_{10}$-$[(da)_{10}$-$(dC)_{10}$-$(da)_{10}]$. The PNA-DNA hybrid was prepared as shown in Figure 1. (A) The hybrid was loaded on a size-exclusion G-4000-DNA-PW HPLC column (7.8 × 300 mm) from Toso (Japan) and was isicratically eluted in 20 mM Tris-Acetate, pH 8.0. The elution was followed at 260 (blue curve) and 495 nm (red curve). The inset shows a UV/VIS spectrum of the eluted hybrid. (B) 20 μL of the hybrid solution were loaded on a 1.5% agarose gel (7 × 7 cm). Electrophoresis was conducted at 4°C and 130 V for 40 min in TEA buffer. The gel was stained with ethidium bromide.

3.3 Synthesis of Conjugates between the DNA-PNA Hybrid and AgNPs

The phosphorothioated residues can covalently anchor the hybrid ends to the surface of silver and gold particles [6, 17], yielding nanoparticle-DNA-PNA conjugates. We have shown that incubation of $(dA)_{10}$-coated AgNPs with the phosphorothioate-functionalized hybrid, $(pG)_{10}$-$[(da)_{10}$-$(dC)_{10}$-$(da)_{10}]$ as described in Experimental Procedures section yielded structures that move through the gel slower than individual AgNPs. A new yellow band (marked by arrow "2", see lane 3 in Figure 3) is clearly seen in the gel image. Incubation of AgNPs with the hybrid not containing phosphorotioated residues, $(pG)_{10}$-$(dC)_{10}$, under identical conditions did not yield the above band (see Figure 3, lane 2). Slices corresponding to bands "1" and "2" (see Figure 3) were cut out of the gel with a razor blade. The molecules were electroeluted from the slices and characterized by TEM as described in Section 2.

TEM analysis (see Figure 4) shows that the waste majority of structures electroeluted from slices marked with arrows "1" and "2" composed of 1 and

Figure 3 Electrophoretic separation of AgNPs-DNA-PNA conjugates. AgNPs (OD at 400 nm is equal to 800) (lane 1) were incubated with 0.8 μM $(pG)_{10}$-$(dC)_{10}$ (lane 2) and 0.8 μM $(pG)_{10}$-[$(da)_{10}$-$(dC)_{10}$-$(da)_{10}$] (lane 3) in: 5 mM K-Pi, pH 7.5 and 0.1 M NaCl at ambient temperature for 24 h. 10 μL of each sample was loaded onto a 1.5% agarose gel and electrophoresed at 4°C and 130 V for 40 min.

2 nanoparticles respectively. Structures electroeluted from the yellow area above band "2" (see lane 3 in Figure 4) were composed of 3–4 particles (data not presented). The interparticle separation distance in the conjugates is approximately equal to 3–4 nm (see inset in the right panel of Figure 4B) that corresponds nicely with the length of a ds-PNA-DNA linker.

As seen in Figure 5 the absorption spectrum of the AgNP-PNA-DNA conjugate is broader than that of the particles not connected to each other. The observed broadening of Ag surface plasmon band around 400 nm reflects the coupling between closely spaced AgNPs in the structure.

We have shown that the conjugates are stable and do not dissociate into single particles at low ionic strength. Incubation for 2 days at ambient temperature in distillate water produced no effect either on the absorption spectrum or on the mobility of the conjugate band through the gel (data not shown).

4 Discussion

We have synthesized a PNA-DNA hybrid composed of a PNA strand, $(pG)_{10}$ and a DNA strand functionalized with phosphorithioated adenine residues on

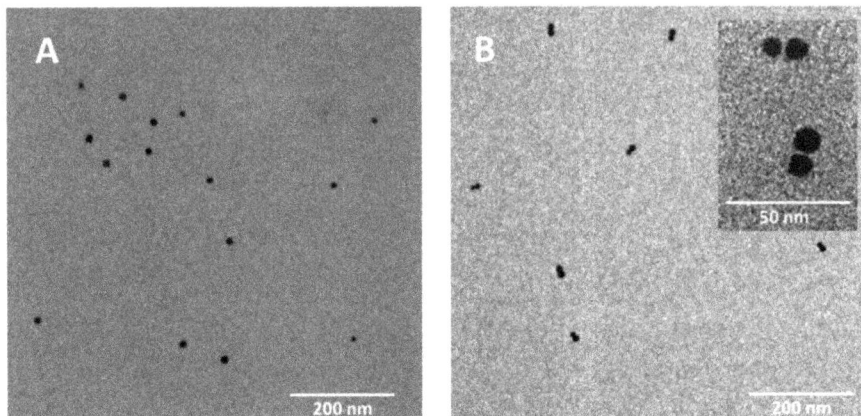

Figure 4 TEM images of AgNPs and the AgNPs-PNA-DNA structures. Molecules electroe-luted from gel slices corresponding to bands "1" (A) and "2" (B) (see Figure 3), were deposited on 400 mesh copper carbon grids and visualized by TEM. The inset is an enlarged image of one of the structures.

either side of the DNA oligonucleotide, $(da)_{10}$-$(dC)_{10}$-$(da)_{10}$. The hybrid is stable under low ionic strength conditions when the two strands of canonical DNA dissociate from each other.

We have demonstrated that silver nanoparticles in contrast to gold ones can be stabilized by regular adenosine oligonucleotides (not containing thiols, phosphorothioates or amines). Incubation of AgNPs with a great molar excess of oligonucleotides composed of either 10 adenosine, cytosine or guanine bases, yielded stable nanoparticles that do not precipitate at salt concentrations up to 300 mM. In contrast, incubation with $(dT)_{10}2$ did not affect the particles stability. This result corresponds well with earlier observations, showing that thymine bases interact with the surface of silver [18] and gold [19, 20] particles weaker than other DNA bases. The week interaction of thymine thus may due to the absence of an exocyclic amine group in this base.

We have shown (see Figures 4 and 5) that incubation of $(dA)_{10}$-coated AgNPs with PNA-DNA flanked with runs of ten phosphorothioated adenine-bases on both ends of the hybrid results in the formation of structures containing two nanoparticles. The absorption spectrum of the dimer is broader than that of AgNPs not connected to each other. It is well known that plasmons of closely spaced nanoparticles are strongly coupled [21, 22]. The spec-

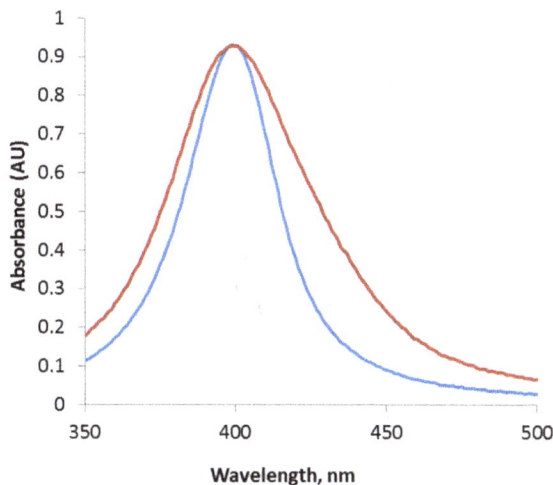

Figure 5 Normalized absorption spectra of AgNPs (blue curve) and AgNP-PNA-DNA conjugates (red curve) electroeluted from a gel slice corresponding to band "2" (see Figure 3).

trum broadening (see Figure 5) thus reflects the presence of electromagnetic coupling between plasmons of AgNPs in the dimer.

The ability of DNA to conduct electrical current depends critically on π-π interaction between stacked Watson–Crick base pairs in the ds-polymer. Even partial strands separation, can strongly reduce, if not completely abolish, conductivity of the DNA molecule. The separation of DNA strands under conditions of low ionic strength might be the reason for poor conductivity reported earlier [23]. PNA-DNA hybrids lack the above disadvantage. The molecule adopts a double helical conformation that is very similar with respect to π-π interactions to a native ds-DNA even in pure water. Attachment of two 15 nm silver particles to both ends of the hybrid is also an advantage that makes possible to place the conjugate in between electrodes separated by 20–30 nm and measure conductivity of very short ($\sim$3 nm in our case) DNA-PNA molecules. We believe that conjugates of nanoparticles with PNA-DNA will possess interesting conductive and optical properties leading to the development of new DNA-based nano-conductors, semiconductors and electro-optical switches.

Acknowledgments

This work was supported by the Israel Science Foundation, 172/10. M.A. acknowledges fellowship support from the Nerviano Medical Sciences.

References

[1] M. R. Jones, K. D. Osberg, R. J. Macfarlane, M. R. Langille, C. A. Mirkin, Chem. Rev., 111, 3736–3827 (2011).

[2] A. P. Alivisatos, K. P. Johnsson , X. G. Peng, T. E. Wilson, C. J. Loweth, M. P. Bruchez, P. G. Schultz, Nature, 382, 609-611 (1996).

[3] C. J. Loweth, W. B. Caldwell, X. G. Peng, A. P. Alivisatos , P. G. Schultz, Chem. Int. Ed., 38, 1808–1812 (1999).

[4] S. Bidault , F. J. G.de Abajo, A. Polman, JACS, 130, 2750–2751 (2008).

[5] A. J. Mastroianni, S. A. Claridge, A. P. Alivisatos, JACS, 131, 8455–8459 (2009).

[6] I. Lubitz, A. Kotlyar, Bioconjug. Chem., 22, 2043–2047 (2011).

[7] N. Borovok, E. Gillon, A. Kotlyar, Bioconjug. Chem., 23, 916–922 (2012).

[8] D. Nykypanchuk, M. M. Maye, D. van der Lelie, O. Gang, Nature, 451, 549–552 (2007).

[9] S. Y. Park, A. K. R. Lytton-Jean, B. Lee S. Weigand, G. C. Schatz, C. A. Mirkin, Nature, 451, 553–556 (2008).

[10] D. Porath, G. Cuniberti, R. Di Felice, Top Curr. Chem., 237, 183–228 (2004).

[11] R. G. Endres, D. L. Cox, R. R. P. Singh, Rev. Mod. Phys., 76, 195–214 (2004).

[12] P. E. Nielsen, M. Egholm, R. H. Berg, O. Buchardt, Science, 254, 1497–1500 (1991).

[13] M. Egholm, O. Buchardt, L. Christensen, C. Behrens, S. M. Freier, D. A. Driver, R. H. Berg, S. K. Kim, B. Norden, P. E. Nielsen, Nature, 365, 566–568 (1993).

[14] C. R. Cantor, M. M. Warshaw, H. Shapiro, Biopolymers, 9, 1059–1077 (1970).

[15] L. Christensen, R. Fitzpatrick, B. Gildea, K. H. Petersen, H. F. Hansen, T. Koch, M. Egholm, O. Buchardt, P. E. Nielsen, J. Pept. Sci., 1, 175–183 (2004).

[16] J. Yguerabide, E. E. Yguerabide, Anal. Biochem., 262, 137–156 (1998).

[17] I. Lubitz, A. Kotlyar, Bioconjug. Chem., 22, 482–487 (2011).

[18] S. Basu, S. Jana, S. Pande, T. Pal, J. Coll. Int. Sci., 321, 288–293 (2008).

[19] A. Gourishankar, S. Shukla, K. N. Ganesh, M. Sastry, J. Am. Chem. Soc, 126, 13186–13187 (2004).

[20] L. M. Demers, M. Ostblom, H. Zhang, N. H. Jang, B. Liedberg, C. A. Mirkin, J. Am. Chem. Soc., 124, 11248–11249 (2002).

[21] S. K. Ghosh, T. Pal, Chem. Rev., 107, 4797–4862 (2007).

[22] N. G. Khlebtsov, L. A. Dykman, J. Quantit. Spectrosc. Radiative Transfer, 111, 1–35 (2010).

[23] R. G. Endres, D. L. Cox, R. R. P. Singh, Rev. Mod. Phys., 76, 195–214, (2004).

Biographies

Gennady Eidelshtein received his B.Sc degree in Biotechnology from Hadassah College, Jerusalem, Israel, in 2009. In 2012 Gennady holds his Master's degree in Materials & Nanotechnology at Tel Aviv University under supervision of Professor Alexander Kotlyar. Today Gennady is Ph.D. student in Alexander Kotlyar's group. His research is focused on novel metal and DNA-based functional nanomaterials.

Shay Halamish was born in Jerusalem, Israel, January 1984. He recieved his B.Sc degree in Biotechnology from Bar Ilan University Israel in 2010. In 2010 Shay started his master's degree in Materials & Nanotechnology at Tel Aviv University under supervision of Professor Alexander Kotlyar. His research is focused on optical properties of metal nanoparticles.

Irit Lubitz was born in Ramat Gan, Israel, January 1977. She received her B.Sc degree in biology from Bar Ilan University, Israel. She holds Master's and Ph.D. degrees in Biochemistry, under supervision of Professor Alexander Kotlyar, from Tel Aviv University. Since 2005 her research has been focused on novel DNA-based functional nanomaterials.

Marcello Anzola was born in Parma, Italy, in 1986. He received his M.Sc degree in Chemistry at Università degli Studi di Parma in 2010. His research was focused on PNAs as diagnostic tools. He currently works at Nerviano Medical Sciences, an Italian research center on oncologic drugs.

Clelia Giannini graduated in Pharmaceutical Chemistry and Technology in 1996 and received her Ph.D. in Bioactive Natural Compound at University of Naples (Italy) in 2001. She joined, as post-doc, the Organic and Industrial Chemistry Department of University of Milan in 2002. She became Assistant Professor in Organic Chemistry in 2005 at the same university. Her research activity is mostly in the fields of the synthesis and structural studies of modified Peptide Nucleic Acid (PNA).

Alexander Kotlyar graduated in Biochemistry and got his Ph.D. in Biochemistry at Moscow State University in 1985. He is a faculty member of Tel Aviv University since 1994. He is an author of about 100 peer reviewed journal papers. His major research interests since 2001 focus around the DNA-based nanostructures. He developed novel methods for enzymatic synthesis of various DNA nanowires and complexes of the wires with metal particles.

Formation of Dimers Composed of a Single Short dsDNA Connecting Two Gold Nanoparticles

Haya Dachlika,[1] Avigail Stern,[1] Dvir Rotem and Danny Porath*

Institute of Chemistry and The Harvey M. Krueger Center for Nanoscience and Nanotechnology, The Hebrew University of Jerusalem, 91904 Jerusalem, Israel
Corresponding author: danny.porath@mail.huji.ac.il

Received 27 November 2012; Accepted 17 December 2012

Abstract

We report synthesis of dimers composed of a *single* short double-stranded (ds)DNA molecule connecting two gold nanoparticles (GNPs). Such structures may be useful for electrical transport measurements through dsDNA molecules and for other research purposes. When the DNA molecules are short with respect to the size of the GNP, gel electrophoresis cannot separate GNPs with different numbers of DNA molecules attached to them. We present two methods to separate GNPs connected to single short thiolated single-stranded (ss)DNA. The separation is performed by hybridizing the DNA/GNP conjugates with long, partially complementary, ssDNA or with complementary ssDNA connected to GNPs of smaller size. The separated GNPs with a single short ssDNA were used to form dimers consisting of GNPs connected by a *single* short dsDNA molecule.

Keywords: DNA, gold nanoparticles, gel electrophoresis.

---

[1] These authors made equal contributions to this manuscript.

Journal of Self-Assembly and Molecular Electronics , Vol. 1, 85–99.

1 Introduction

Many conjugate structures composed of DNA and gold nanoparticles (GNPs) have been reported in the literature. A common method for forming these conjugates is by connecting the DNA strands to the GNP through thiol chemistry, which offers the ability to design complex nanoparticle assemblies [1–3]. Such conjugates have been used for various emerging challenges such as electrical transport measurements [4–6], surface plasmon resonance (SPR) studies [7, 8], molecular nanomachines formation [9] and others. Thanks to their optical properties [10], DNA/GNP conjugates were also applied to the development of applications in biosensing [11, 12], nanobiotechnology [11], nanomedicine [13, 14] and nanoelectronics [15]. For many of these applications it is important to control the exact number of DNA strands attached to the nanoparticles. The ability to separate conjugates of GNPs with a specific number of short DNA strands, as building blocks for 2D and 3D GNP-DNA nano-constructions, e.g. dimers, is challenging and has attracted much scientific interest.

Gel electrophoresis is a common technique for separation of DNA/GNP conjugates because additional DNA molecules connected to the GNP produce a significant shift in the electrophoretic mobility of the conjugates [16, 17]. Gel electrophoresis separation is, however, limited to particles connected to relatively long oligonucleotide strands. The sensitivity and the effectiveness of the method are reduced when the particle size is comparable or larger than the DNA length. Recently, some methods to synthesize GNPs attached to specific numbers of DNA strands have been employed: long DNA strands were first used to get efficient separation with gel electrophoresis and then removed or shortened in order to get GNPs with short DNA strands [18, 19]. Another method used anion-exchange HPLC as a technique for conjugate separation [20].

In this work we present two easy and convenient methods for separation of conjugates of GNPs with specific numbers of *short* oligonucleotide strands. The separation is implemented for and confirmed by the formation of GNP dimers connected by single dsDNA molecules. The separation methods are based on intermediate increase in the variability between the conjugates. This leads to a significant difference in the electrophoretic mobility of the different structures. Characterization of the products was done by transmission electron microscopy (TEM).

Table 1 DNA sequences used in the synthesis of the four DNA/GNP conjugates.

Strand 1	S26_1	SH-CATTAATGCTATGCAGAAAATCTTAG
Strand 2	S26_2	SH-CTAAGATTTTCTGCATAGCATTAATG
Strand 3	RC100_1	TTAATGCTATGCAGAAAATCTTAG(T×76)
Strand 4	RC100_2	AAGATTTTCTGCATAGCATTAATG(A×76)

2 Experimental Procedures

2.1 Generation of GNP/DNA Conjugates

Citrate stabilized GNPs with mean diameters of 5 and 8 nm were purchased (Ted Pella). The GNPs were coated with phosphine ligands (bis(p-sulfonato-phenyl)-phenylphosphine (BSPP) as previously described by Mastroianni et al. [21]. The GNPs were stabilized with 1 mM BSPP by overnight incubation at room temperature. NaCl was then added to the GNP solution until a color change from red to blue was observed (the final NaCl concentration was ~200 mM for 8 nm GNP and ~300 mM for 5 nm GNP). The GNP solution was concentrated by centrifugation at 3000 rpm for 10 minutes at room temperature. The supernatant was removed and the pellet was resuspended in 1 mg ml^{-1} BSPP solution (concentration of 3.14 mM) to ~1 μM. The phosphine coating provides a net negative charge on the GNP surface, thus stabilizing it in high concentrations and in buffer conditions.

5' thiol-C6-modified ssDNA oligonucleotides with 26 bases (sequences presented in Table 1) were purchased (Syntezza). The oligonucleotides were treated with tris(2-carboxyethyl)phosphine hydrochloride (TCEP) at a TCEP:DNA molar ratio of 1000:1 for removal of the protective group on the thiol, and then purified with Bio-Spin 6 columns (Biorad).

GNP/DNA conjugates were formed by incubating the BSPP coated GNPs with the treated DNA at a final concentration of 0.5 μM, DNA:GNP molar ratio of 1:1 overnight at room temperature (25°C) in 100 mM tris buffer (pH 7.4) with 50 mM NaCl.

2.2 Gel Electrophoresis Separation and Characterization

Two methods were used to separate conjugates of 8 nm GNPs with different numbers of 26 bases ssDNA strands. In the first method the DNA strands were conjugated to 8 nm GNPs and hybridized to complementary 26 bases ssDNA strands, which were conjugated to 5 nm GNPs. For the hybridization the two conjugate solutions (5 nm GNP conjugates and 8 nm GNP conjugates) were mixed at a 10:1 molar ratio, respectively, and incubated over night at room temperature. After the hybridization, the obtained structures were separated by gel electrophoresis in 3% MetaPhor agarose (Lonza) with 1X TBE as running buffer at 100 V for 60 min.

In the second method 100 bases ssDNA strands were hybridized to 26 bases ssDNA strands (over 26 bp length), which were already conjugated to the GNPs using thiols. The molar ratio of the hybridized 26bp:100bp conjugates was 1:10 respectively and the incubation was performed overnight at room temperature. The resulting structures were separated by gel electrophoresis in a 3% MetaPhor agarose gel with 1X Tris/Borate/EDTA (TBE) buffer as running buffer for 1.5 h at 100 V. The separation of the 20 nm GNPs was done in a 1.5% MetaPhor agarose gel under the same conditions.

2.3 Dimers Preparation

Conjugates of GNPs with DNA molecules that were prepared by the second method described above were extracted from the gel by electroelution at 100 V for half an hour. Conjugates of GNPs bound to a single ssDNA molecule (mono-conjugates) were concentrated similarly to the concentration procedure described above for BSPP stabilized GNPs. Mono-conjugates with complementary DNA sequences were then mixed, heated to 60°C for 10 min (for dehybridization of the 100 bases strands from the 26 bases strands and their hybridization with the complementary 100 bases strands) and then cooled down to room temperature (for hybridization of the 26 bases strands to the complementary ones). Hybridization of complementary 100 bases strands and complementary 26 bases strands is thermodynamically favored. Following hybridization of the complementary mono-conjugates, the structures were again separated by 3% MetaPhor agarose gel electrophoresis with 1X TBE as the running buffer at 100 V for 60 min. Dimers were then extracted from the gel by electroelution at 100 V for 60 min.

As a control experiment, dimers were also prepared without using any separation method. These dimers were prepared by mixing non-separated 8 nm GNP/DNA conjugate solutions and incubating them over night at room temperature for hybridization.

2.4 Deposition and TEM Imaging

Synthesized dimers were deposited on a 400 square mesh carbon coated copper TEM grid (Electron Microscopy Sciences) and imaged by a Tecnai T12 G2 Spirit TEM (FEI). Deposition was carried out by first treating the grid with 0.1% w/v poly-L-lysine in water for 1 min, washing by floating on a water drop for another minute, and then floating on the sample drop for three minutes. Images were taken with an FEI Eagle 4k, 16 Megapixel CCD camera.

3 Results and Discussion

Gel electrophoresis is often used as a tool in the characterization and puri-fication of GNPs/DNA conjugates [16, 17]. The basic principle relies on the characteristic mobility of objects in a porous matrix, or gel, under an electric field. The mobility (distance/(electric field * time)) depends mainly on the size and the charge of the conjugates [16]. A good separation between GNPs bound to different numbers of DNA molecules is achieved when the bound DNA molecules are much longer than the GNP diameter, e.g., GNPs with a diameter of 8 nm and ssDNA of 86 bases (see Figure 1(a)) [16, 20]. When the DNA molecules are short relative to the size of the GNPs, it is difficult or even impossible to separate DNA/GNP conjugates based on the number of bound DNA molecules. For instance, in the gel shown in Figure 1(b) it is impossible to distinguish between bare GNPs (lane 1) and conjugates of 8 nm GNPs bound to short, 26 bases long, thiolated DNA oligonucleotides (lane 2). Therefore, in order to prepare mono-conjugates of 8 nm GNP with a single 26 bases ssDNA molecule more refined separation methods must be used.

We describe two methods that enable to separate GNPs connected to different numbers of short ssDNA strands. These methods are based on tem-porary hybridization of the short DNA strands that are bound to the GNPs to complementary short ssDNA bound to a smaller GNP or to longer ssDNA strands. In the first method 8 nm GNPs were first bound to 26 bases ssDNA molecules (strand S26_1) and 5 nm GNPs were bound to the complement-ary 26 bases ssDNA molecules (strand S26_2). Then, the two solutions of GNP/DNA conjugates were mixed, hybridized, (with the 5 nm conjugates in a large excess) and separated by gel electrophoresis (Figure 2). Because the small conjugates were in large excess, they were assumed to hybridize with all the DNA strands bound to the 8 nm GNPs. This way the 5 nm GNP conjugates enable separation of the 8 nm GNP conjugates based on the number of DNA molecules (and 5 nm GNPs) bound to them. Therefore, after separation we expect that the vast majority of dimers will constitute 8 nm GNPs bound to a single DNA molecule that is hybridized to a single 5 nm conjugate (forming dimers with a single dsDNA molecule). Figure 2

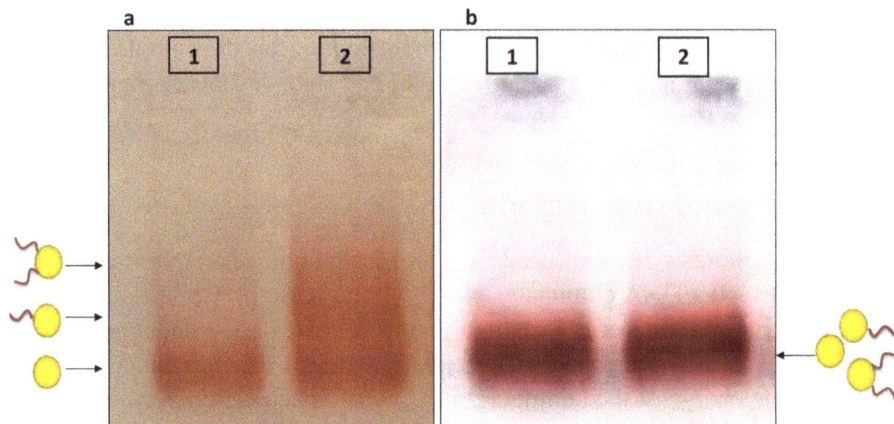

Figure 1 Effect of DNA length on separation of DNA/GNP conjugates with gel electrophoresis. 8 nm GNPs were attached to thiolated DNA molecules as described in Section 2. (a) Lane 1 contained a control sample of 8 nm GNPs with no DNA. Lane 2 contained conjugates with 86 bases DNA molecules. The conjugate mixtures were separated into bands, each containing a different number of DNA strands as indicated by the illustrations next to the arrows (adapted from [1]). (b) Lane 1 contained a control sample of 8 nm GNPs with no DNA. Lane 2 contains conjugates of 8 nm GNPs with 26 bases DNA molecules (strand S26_1). The conjugate mixtures were not separated into bands in this gel.

shows a gel in which such a separation was carried out. Lanes 1 and 2 are control lanes: lane 1 contains 5 nm GNPs with no DNA bound to them; lane 2 contains 8 nm GNPs with no DNA bound to them. Lane 3 contains the hybridized mixture of the two GNPs/DNA conjugates, which was separated into bands. These bands contain, from bottom up, 5 nm GNPs, 8 nm GNPs, 8 nm+5 nm dimers. The following bands contain higher oligomeric structures, as indicated by the illustrations next to the arrows on the right. In order to visualize and check the formation of dimer structures, the third band from the bottom (framed), which was expected to contain dimers, was extracted from the gel for further analysis by TEM imaging (Figure 3). The majority of the observed structures are of 5–8 nm GNPs dimers. These results demonstrate that the hybridization of additional structures in large excess can assist separation between GNP/DNA conjugates based on the number of short ssDNA strands that they contain.

In the second method, we hybridized the 8 nm GNP/26 bases ssDNA (strands S26_1 and S26_2) conjugates with long ssDNA molecules (100 bases ssDNA, strands RC100_1 and RC100_2) with a 26 bp overlap, as illustrated

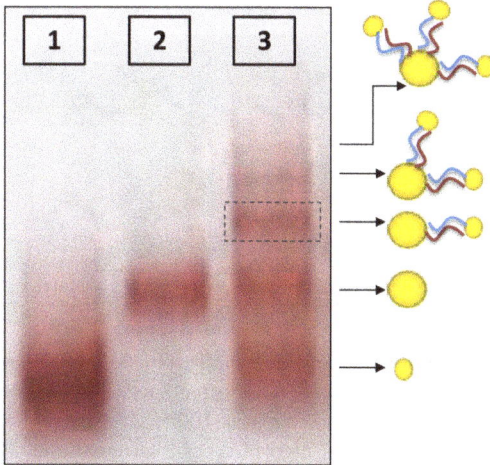

Figure 2 Electrophoretic separation of 8 nm GNP/26 bases DNA (S26_1) conjugates through binding to 5 nm GNP/26 bases DNA (S26_2) conjugates. Lane 1 contains a control sample of 5 nm GNPs with no DNA. Lane 2 contains a control sample of 8 nm GNPs with no DNA. Lane 3 contains the hybridized mixture of the two types of GNP/DNA conjugates with complementary 26 base DNA strands. The conjugates mixture was separated into bands indicating that the hybridization produced structures of 8 and 5 nm GNPs that are connected by dsDNA as illustrated next to the arrows. The structures in the band in lane 3 marked by the dashed frame were extracted from the gel and imaged by TEM (see Figure 3).

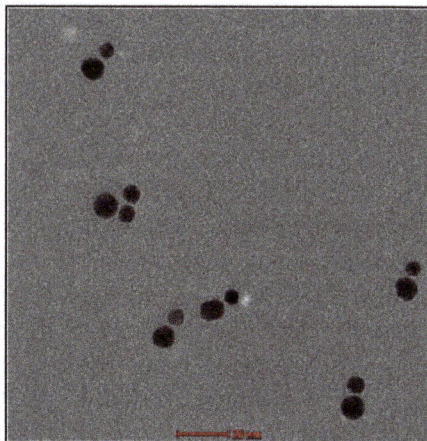

Figure 3 TEM image of 8 and 5 nm GNP dimer structures that are connected by 26 bases dsDNA. The structures were extracted from the gel band marked by the frame in Figure 2. The majority of the structures imaged were dimers of 8 nm GNPs with 5 nm GNPs. Scale bar 20 nm.

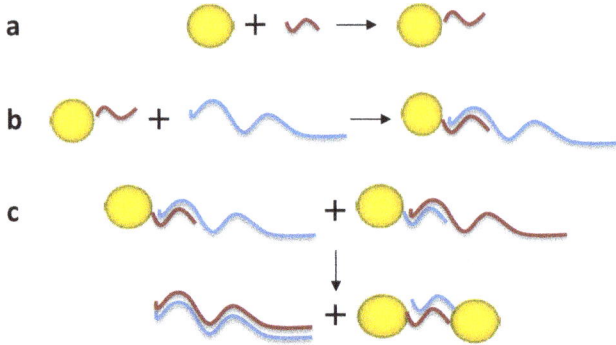

Figure 4 Illustration of the procedure for the synthesis of dimers composed of GNPs connected by a single dsDNA molecule. (a) Conjugation: 8 nm GNPs were attached to 26 bases ssDNA. (b) Hybridization with a complementary 100 bases ssDNA over the 26 bp complementary part. Gel electrophoresis was used in order to separate the GNP/DNA conjugates bound to a single ssDNA molecule. (c) The GNP/DNA conjugates were extracted from the gel and mono-conjugates with the complementary short DNA molecules were mixed. By heating and then cooling down the solution, the complementary 100 bases strands form dsDNA and the GNPs that were bound to the short DNA molecule formed dimers. Following hybridization, the structures were again separated by gel electrophoresis (see Figure 5).

in Figure 4. Again, because the long ssDNA molecules were in a large excess one may assume that they hybridized to all the short ssDNA molecules bound to the 8 nm GNPs. Similarly to the previous method, this hybridization enabled conjugates separation based on the number of 26 bases ssDNA molecules that are bound to them, which are now extended by the 100 bases strands.

Figure 5 shows a gel in which such a separation was carried out. Lane 1 contains 8 nm GNPs with no bound DNA. The other two lanes contain conjugates of 8 nm GNPs with 26 base ssDNA after their hybridization with 100 base DNA molecules. Each one of these lanes contained conjugates with one of the two complementary 26 bases ssDNA strands (strand S26_1 hybridized with RC100_1 and strand S26_2 hybridized with RC100_2, respectively). The conjugate mixtures were separated in this gel into bands, each containing a different number of DNA strands bound to the GNP, as illustrated next to the arrows on the right. This clearly shows that the extension with the longer DNA strands influence the conjugates' mobility and enables the desired separation. The inset in Figure 5 demonstrates separation of 20 nm GNPs attached to 26 bases ssDNA (strands S26_1) after their hybridization with 100 bases ssDNA strands (strand RC100_1). The right lane (that contains

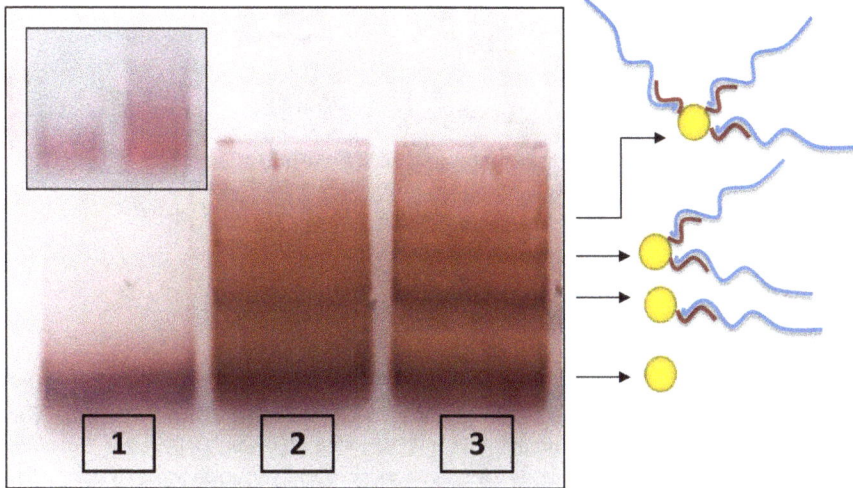

Figure 5 Electrophoretic separation of GNP/26 bases ssDNA hybridized with complementary long DNA molecules. Lane 1 contains a control sample of 8 nm GNPs with no DNA. The other lanes contain conjugates of 8 nm GNPs with each of the two complementary 26 bases ssDNA molecules after their hybridization with 100 base DNA molecules (lane 2 contains 8 nm GNPs attached to S26_1 hybridized to RC100_1, lane 3 contains 8 nm GNPs attached to S26_2 hybridized to RC100_2). The conjugate mixtures were separated in this gel into bands, each containing a different number of DNA strands as indicated by the illustrations next to the arrows to the right. Inset: the left lane contains conjugates of 20 nm GNPs with 26 bases ssDNA molecules (S26_1). The right lane contains conjugates of 20 nm GNPs with 26 bases ssDNA molecules (S26_1) after their hybridization with 100 bases ssDNA molecules (RC100_1).

the 100 bases strands) was separated in this gel into bands. Those results demonstrate that the separation method can be affective and useful also for larger GNPs. Comparison of the separation obtained in the lanes that contain 10 nm (the two right lanes in Figure 5) and 20 nm (the right lane in the inset in Figure 5) GNPs shows, however, that the separation efficiency drops for larger GNPs for a given DNA length.

GNP/DNA mono-conjugates were extracted from the second band from the bottom in lanes 2 and 3 in the gel, and mono-conjugates with complementary sequences were mixed. The 100 bases ssDNA strands were de-hybridized from the conjugates by heating, and then the solution was cooled down. During the cooling, the 100 bases ssDNA molecules hybridize to each other at relatively high temperatures, before the temperature is low enough for them to re-hybridize to the short conjugates. Therefore, when the temperature is

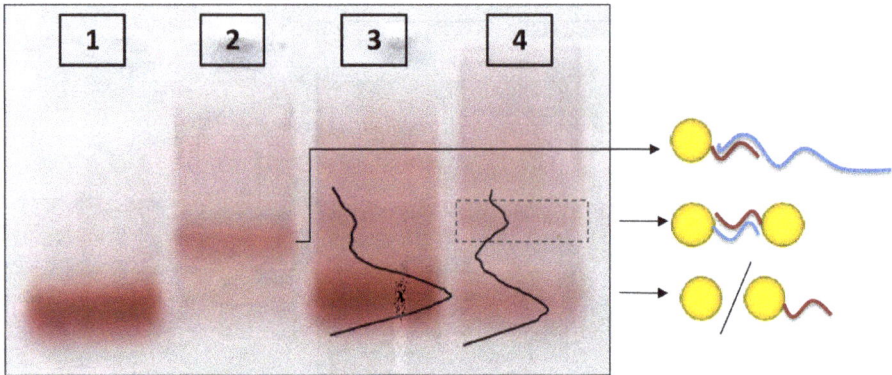

Figure 6 Separation of DNA/GNP conjugates by gel electrophoresis. Lanes 1–3 contain control samples: Lane 1 contains a sample of 8 nm GNPs with no bound DNA. Lane 2 contains conjugates of 8 nm GNPs with 26 bp DNA sequences (S26_1) hybridized to 100 bases ssDNA (RC100_1); lane 3 contains a hybridized mixture of 8 nm GNPs that are bound to 26 bases complementary ssDNA molecules (8 nm GNPs attached to S26_1 hybridized to 8 nm GNPs attached to S26_2). In this case, the conjugates might contain varying numbers of DNA molecules. Lane 4 contains a hybridized mixture of two complementary mono-conjugates (8 nm GNPs attached to single S26_1 and to 8 nm GNPs attached to single S26_2) after they were separated by hybridization to long DNA (RC100_1 and RC 100_2 respectively) so it is verified that only one dsDNA connects the GNPs in each dimer. The arrows point to illustrations of the expected structures.

further decreased the 100 bases strands are not in ssDNA form anymore and not available to rebind to the short 26 bases ssDNA molecules that hybridize to each other to form dimers, as illustrated in Figure 4 and demonstrated in Figure 6. Figure 6 shows the gel in which samples were separated after hybridization of the mono-conjugates. Lanes 1-3 are control lanes. Lane 1 contains a control sample of 8 nm GNPs with no bound DNA; lane 2 contains conjugates of 8 nm GNPs with 26 bp ssDNA (S26_1) hybridized to one 100 bases ssDNA (RC100_1) (similar to the second band in lane 2 in Figure 5) and lane 3 contains a hybridized mixture of 8 nm GNP/26 base ssDNA conjugates, which had not been separated prior to hybridization. Lane 4 contains a hybridized mixture of two complementary mono-conjugates after they were separated by hybridization to 100 bases ssDNA molecules. In lanes 3 and 4 a separation to bands is observed. The bands in these lanes contain from bottom up: mono-conjugates (mixed with bare GNPs), dimers and higher oligomeric structures. In lane 4 there is no band parallel to the one in lane 2 (mono-conjugates of 8 nm GNPs bound to 26 bp DNA strand that hybridized to

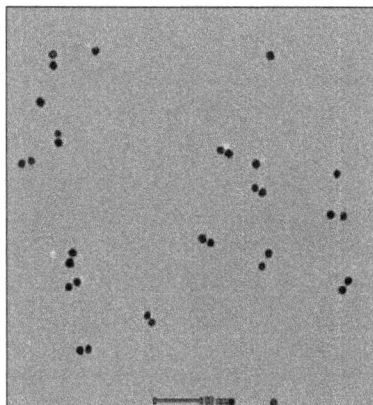

Figure 7 TEM image of the synthesized GNP dimers after using the separation method described in the text. The structures were extracted from the gel band indicated by the frame in Figure 6, lane 4. Most of the particles are assembled into dimers. Scale bar 50 nm.

100 bases complementary DNA). This indicates clearly that the hybridization between two 100 bases DNA strands can effectively release the 26 bases ssDNA bound to the GNP so they are free to form dimers. Note that the total "color" in lane 3 that did not pass a first separation is stronger, as it overall contains more GNPs, but the relative intensity of the dimer band (second from bottom) in lane 4 compared to the bare GNP band (first band) is stronger than in lane 3 since after the first separation the relative amount of dimers is larger.

The bands that were expected to contain dimers were extracted from the gel for further analysis by TEM. TEM characterization of the dimer band in lane 4 is presented in Figure 7. The TEM imaging demonstrates that dimers were indeed formed and that the majority of the particles belong to dimers. Though the second band from the bottom in both lanes 3 and 4 was shown (by TEM characterization) to contain dimer structures, one must note that while the GNPs forming the dimers in lane 3 may be attached by more than one dsDNA molecule and may have additional ssDNA molecules bound to them, the dimers in lane 4 have only one dsDNA molecule attaching the GNPs and no additional ssDNA molecules. This property cannot be characterized either by gel electrophoresis or by TEM, but it is essential and stems from our synthesis procedure.

4 Conclusions

When GNPs and DNA are mixed in 1:1 ratio, at least some of the GNPs are connected to more than one DNA strand. Therefore, dimers directly formed by mixing these conjugates may be connected by more than a single dsDNA, and may have additional ssDNA molecules bound to them. When the DNA molecule is long, gel electrophoresis can be used for separation between GNPs attached to different numbers of DNA strands. When short DNA strands are connected to the GNPs, however, the DNA/GNPs mono-conjugates cannot be directly separated by gel electrophoresis. We show here two simple methods by which one can efficiently separate conjugates bearing different numbers of short DNA strands. These two methods provide the requested resolution in gel electrophoresis and enable the isolation of mono-conjugates containing short DNA strands.

Each one of the two presented methods has its own advantages. In the first method the DNA/GNP conjugates were hybridized to complementary ssDNA molecules connected to smaller GNPs. This method allows validation of the process at intermediate stages. After the separation stage, for example, the dimers (formed from monoconjugates hybridized to only one complementary ssDNA attached to a small GNP) can be imaged by TEM. In the second method the DNA/GNP conjugates were hybridized with long complementary ssDNA. This method enables easy and convenient release of the extending strand from the conjugates for further use of the mono-conjugates. In our work, for example, they were used for the formation of dimers composed of two equal sized GNPs connected by a single dsDNA.

The two methods we demonstrate here for separation of 8 nm GNPs conjugated to 26 base ssDNA can be implemented also for larger GNPs. As the GNP diameter increases the separation efficiency for a given DNA length drops (as demonstrated in the inset in Figure 5). Effective separation can be maintained, however, by simply increasing the length of attached DNA by hybridization with longer complementary DNA strands. In addition to dimers formation, nanoparticles with a discrete number of DNA molecules can be used as building blocks for a variety of nanostructures. The methods we present here may be implemented for preparation of such nanostructures.

Acknowledgments

This work was supported by the European Commission through FP6-IST grant for Future & Emerging Technologies 'DNA-based Nanodevices' (FP6-

029192); the ESF COST MP0802; the Israel Science Foundation (grant no. 1145/10); BSF grant 2006422; the French Ministry of External Affairs, the Israeli-Palestinian Science Organization and Friends of IPSO, USA (with funds donated by the Meyer Foundation); Academia Sinica and the Hebrew University of Jerusalem joint research program in Nanoscience and Nanotechnology. HD thanks the Klein foundation for MSc scholarships.

References

[1] A. Stern, D. Rotem, I. Popov, D. Porath, Journal of Physics: Condensed Matter, 24(16), 164203 (2012).

[2] C. J. Loweth, W. B. Caldwell, X. Peng, A. P. Alivisatos, P. G. Schultz, Angewandte Chemie International Edition, 38(12), 1808–1812 (1999).

[3] J. Zheng, P. E. Constantinou, C. Micheel, A. P. Alivisatos, R. A. Kiehl, N. C. Seeman, Nano Letters, 6(7), 1502–1504 (2006).

[4] H. Cohen, C. Nogues, R. Naaman, D. Porath, Proceedings of the National Academy of Sciences of the United States of America, 102(33), 11589–11593 (2005).

[5] H. Cohen, C. Nogues, D. Ullien, S. Daube, R. Naaman, D. Porath, Faraday Discuss., 131, 367–376 (2005).

[6] D. Ullien, H. Cohen, D. Porath, Nanotechnology, 18(42), 424015 (2007).

[7] A. Kuzyk, R. Schreiber, Z. Fan, G. Pardatscher, E. M. Roller, A. Högele, F. C. Simmel, A. O. Govorov, T. Liedl, Nature, 483(7389), 311–314 (2012).

[8] J. Wirth, F. Garwe, G. Hähnel, A. Csaki, N. Jahr, O. Stranik, W. Paa, W. Fritzsche, Nano Letters, 11(4), 1505–1511 (2011).

[9] T. Song, H. Liang, Journal of the American Chemical Society, 134(26), 10803–10806 (2012).

[10] J. J. Storhoff, A. A. Lazarides, R. C. Mucic, C. A. Mirkin, R. L. Letsinger, G. C. Schatz, Journal of the American Chemical Society, 122(19), 4640–4650 (2000).

[11] J. Liu, Y. Lu, Journal of the American Chemical Society, 125(22), 6642–6643 (2003).

[12] W. Zhao, C. Yao, X. Luo, L. Lin, I. Hsing, Electrophoresis, 33(8), 1288–1291 (2012).

[13] D. A. Giljohann, D. S. Seferos, W. L. Daniel, M. D. Massich, P. C. Patel, C. A. Mirkin, Angewandte Chemie International Edition, 49(19), 3280–3294 (2010).

[14] J. Xue, L. Shan, H. Chen, Y. Li, H. Zhu, D. Deng, Z. Qian, S. Achilefu, Y. Gu, Biosensors and Bioelectronics (2012).

[15] C. L. Choi, A. P. Alivisatos, Annual Review of Physical Chemistry, 61, 369–389 (2010).

[16] D. Zanchet, C. M. Micheel, W. J. Parak, D. Gerion, A. P. Alivisatos, Nano Letters, 1(1),32–35 (2001).

[17] W. Zhao, I. M. Hsing, Chemical Communications, 46(8), 1314–1316 (2010).

[18] N. Borovok, E. Gillon, A. Kotlyar, Bioconjugate Chemistry, 23(5), 916–922 (2012).

[19] M. P. Busson, B. Rolly, B. Stout, N. Bonod, E. Larquet, A. Polman, S. Bidault, Nano Letters, 11(11), 5060–5065 (2011).

[20] S. A. Claridge, H. W. Liang, S. R. Basu, J. M. J. Fréchet, A. P. Alivisatos, Nano Letters, 8(4), 1202–1206 (2008).

[21] A. J. Mastroianni, S. A. Claridge, A. P. Alivisatos, Journal of the American Chemical Society, 131(24), 8455–8459 (2009).

Biographies

Haya Dachlika got her B.Sc degree in Chemistry and Biology from The Hebrew University of Jerusalem, Israel in 2010. In 2011 Haya started her master's degree in Chemistry at The Hebrew University of Jerusalem under supervision of Professor Danny Porath. Her research is focused on novel DNA-based functionalized metal nanoparticles.

Avigail Stern got her B.Sc degree in Biology and Chemistry from the Hebrew University of Jerusalem, Israel, in 2010. Avigail is pursuing her Ph.D. in Chemistry at the Hebrew University of Jerusalem under the supervision of Professor Danny Porath. Avigail's research is focused on charge transport measurements through DNA molecules.

Dvir Rotem got his Ph.D. in Brain and Behavioral Sciences from the Hebrew University of Jerusalem. He joined, as a post-doc, the Bayley group at the Chemistry Department at The University of Oxford, UK in 2005. Since 2011 he is a research associate in the Institute of Chemistry in the Hebrew University of Jerusalem and works as the laboratory manager of Professor Danny Porath's group.

Danny Porath got his Ph.D. in Physics at the Hebrew University of Jerusalem in 1997. Pioneered electrical transport measurements in single DNA molecules as a postdoc in the group of Professor Cees Dekker at Delft University of Technology. A faculty member at the Hebrew University of Jerusalem since 2001. His major research interest is implementation of bio-molecules in molecular electronics devices in general and understanding the electrical charge transport mechanisms in DNA in particular.

Self-Assembled DNA-Based Structures for Nanoelectronics

Veikko Linko[1] and J. Jussi Toppari[2]

[1]Physics Department, Walter Schottky Institute, Technische Universität München,
85748 Garching near Munich, Germany; e-mail: veikko.linko@tum.de
[2]Department of Physics, Nanoscience Center, University of Jyväskylä, P.O. Box 35,
40014 Jyväskylä, Finland; e-mail: j.jussi.toppari@jyu.fi

Received 16 November 2012; Accepted 17 December 2012

Abstract

Recent developments in structural DNA nanotechnology have made complex and spatially exactly controlled self-assembled DNA nanoarchitectures widely accessible. The available methods enable large variety of different possible shapes combined with the possibility of using DNA structures as templates for high-resolution patterning of nano-objects, thus opening up various opportunities for diverse nanotechnological applications. These DNA motifs possess enormous possibilities to be exploited in realization of molecular scale sensors and electronic devices, and thus, could enable further miniaturization of electronics. However, there are arguably two main issues on making use of DNA-based electronics: (1) incorporation of individual DNA designs into larger extrinsic systems is rather challenging, and (2) electrical properties of DNA molecules and the utilizable DNA templates themselves, are not yet fully understood. This review focuses on the above mentioned issues and also briefly summarizes the potential applications of DNA-based electronic devices.

Keywords: Self-assembly, DNA nanostructures, electrical conductivity of DNA, carbon nanotubes, nanoparticles.

Journal of Self-Assembly and Molecular Electronics, Vol. 1, 101–124.

1 Structural DNA Nanotechnology

Since Nadrian Seeman's pioneering work in the beginning of 1980s [1], DNA has been considered as a promising material for nanoscale constructions due to its superior self-assembly characteristics, small size and suitable mechanical properties. The highly specific and predictable Watson-Crick base pairing of complementary base sequences of single-stranded DNA (ssDNA) molecules can be utilized in programming desired double-stranded DNA (dsDNA)-based motifs, which can be further assembled into larger and more complex structures. These DNA structures can also serve as templates for other nanoscale objects.

During recent decades, numerous different DNA structures have been introduced. The very first ones were based on flexible branched junctions [2], but the next generation of structures were already quite complex, comprised of more rigid motifs, such as *double crossover* (DX) and *triple crossover* (TX) tiles [3–6], where parallel DNA helices (two in DX and three in TX) are connected to each other via two strand-exchange points, i.e. crossovers. Later on, it was realized that structures could be formed also by using a long scaffold strand combined with shorter DNA fragments [7, 8]. In 2006, Paul Rothemund presented the DNA origami method [9], based on folding a long single-stranded scaffold strand into a desired shape with the help of a set of ssDNA staples (short oligonucleotides). This robust high-yield method has then been extended to three-dimensional shapes [10], resulting in the structures with stress [11] and complex curvatures [12, 13]. In addition, there exist efficient tools, which can help in designing and simulating the origami shapes [14–16]. The rapid progress and the future challenges of the field have been reviewed in [17, 18]. Very recently, a rapid folding of complex 3D origamis, with yields approaching 100%, has been introduced [19], as well as scaffold-free 2D and 3D architectures, which can act as molecular canvases for creating a huge number of distinct arbitrary shapes, with a fair yield [20, 21]. Moreover, it has been shown that the core of a densely packed origami can have a high-degree of structural order [22], thus supporting the idea of complex, high-resolution platforms for diverse applications. Therefore, all these recent achievements truly expand the possibilities in designing custom, spatially well controlled structures at even subnanometer scale.

The novel DNA designs open up opportunities in many distinct research fields, since the structures can be almost arbitrarily patterned with other nanoscale components such as carbon nanotubes [23], proteins [24, 25], metallic nanoparticles [26–29], and quantum dots [30]; not to mention that the structures could also be completely metallized [31–33]. With the help of a DNA template, the placement and orientation of individual molecules or larger molecular assemblies becomes possible, in principle with an accuracy of a single base pair (0.34 nm height, 2 nm helix diameter). For molecular electronics, it means that DNA architectures could serve as versatile molecular scale circuit boards, enabling fabrication of sophisticated nanodevices well below the 22 nm feature size – the next goal of semiconductor industry [34]. Besides the implicit high-resolution, these methods exploit parallel self-assembly processes and could thus provide cheaper and faster way to fabricate nanoscale devices in comparison to the standard top-down-based methods, and thus offer major advantages for the miniaturization of electronics [35,36].

But could we really utilize DNA molecules as circuit boards? Or could even a DNA template itself behave as a conductor? What is the influence of the environment to a behavior of the fragile DNA? Understanding of electrical properties of DNA is important not only for molecular electronics, but also in a field of organic devices [37], medicine and cancer therapy as well as in investigation of genetic mutations and especially in biological sensing [38–40].

2 DNA in Electronics

2.1 Electrical Properties of Double-Stranded DNA

In 1962, it was first time suggested that dsDNA molecules could conduct electricity due to the overlapping π-orbitals of adjacent bases on the base pair stack [41]. Since then, and in particular during recent twenty years, huge amount of theoretical and experimental articles about DNA conductivity have been published: In 90's the actual charge transfer (CT) from one base to another along the dsDNA helix was proven by chemical approach, within an aqueous buffer, by utilizing modified bases acting as a donor and acceptor while monitoring the quenching of the acceptor fluorescence after triggering the donor [42–44]. Since that there have been variety of studies about the DNA CT processes yielding a mixture of conclusions; the variation being mostly due to the differences in coupling of the donor and the acceptor within the base pair stack [45]. Usually CT studies have covered short distances, but also long range CT has been recently reported [40]. After the promising results based on chemical approach published in 1990s, a large variety of studies with physical approaches, i.e., directly measuring the conductivity of dsDNA, soon followed with contradicting results: insulating [31], ohmic [46], semiconducting [47], and even superconducting [48] properties have been reported. The conductivity can be sequence- and mismatch-dependent [40, 49], and it can also be a combination of nucleotide, backbone, and ion-based conductances [50]. Wide range of distinct and controversial results of dsDNA conductivity and the proposed conductance mechanisms (electronic coupling between π-orbitals of neighboring base pairs [41], tunneling or thermally induced hopping [45]) can be found in [51–53].

There also exist several factors that need to be taken into account in a measurement setup but are in most cases highly non-trivial to control. In many physical experiments the contacts between electrodes and DNA play a crucial role [54], while ensuring a proper electrical contact at a single molecule level is extremely difficult [55]. Further, even if the proper contacts could be realized, various environmental factors, e.g. humidity, can have an influence on the conformation of DNA [56], and also to the conductivity [57, 58]. At the high humidity levels the adsorbed and ionized water molecules surrounding the dsDNA can act as charge carriers [59, 60], or at higher frequencies the conductivity can be ascribed to relaxational losses of the surrounding water dipoles [61]. On the other hand, if electrical measurements are performed in a vacuum chamber, the DNA molecule should completely dehydrate resulting in an unknown conformation. In addition, the type of counter-ions and the salt concentration are known to have a large impact to the secondary structure of dsDNA, and ions can also diffuse and migrate along DNA, thus enhancing an ionic conductivity. Yet, the charge of a DNA molecule affecting to the amount of the counter-ions, depends on the dissociation of the phosphate groups and therefore on pH. The observed conductivity is also dependent on the measurement geometry (DNA lying on the substrate vs. freely hanging geometry) and the type of the substrate used, since the conformation of DNA also depends on the interaction between DNA and the substrate in question [62].

Nevertheless, today it seems to be quite clear that a really long, completely unmodified dsDNA molecule itself does not have high enough conductance for serving as an electrical building block or a wire, and equally, it is not sturdy enough to be used in electronic devices. Yet, the conductance mechanisms of DNA are not fully revealed and the topic of the conductivity of the dsDNA still remains highly controversial. Hence, it is not very well known whether other more complex forms of DNA could provide better properties for electronics, or if relatively rigid DNA constructs or some particular parts of them, could conduct electricity if appropriately designed. There already exist several studies on these issues, of which the former being briefly discussed in the next section. The conductivity of dsDNA-based nanostuctures, being one of the main topics of this review, will be reviewed in more detail in later sections to complete an overall picture of the status of the field.

2.2 Other Linear DNA Conformations

Since the conductivity of a plain long dsDNA molecule has been shown not to be sufficient for electronics, many other DNA conformations and derivates such as metallo-DNAs [63–65] or G-wires [66–69] have been studied. The metallo-DNA (M-DNA) is a derivative of dsDNA in which metal ions are incorporated between the bases by replacing the amino protons of guanine and thymine on each of the base pairs at high pH. The metal ions couple the energy levels of the adjacent bases and lower the energy gap, thus enhancing the conductivity of the dsDNA [65]. This enhanced conductivity has been observed already for Zn/M-DNA [63]. However, the conductivity is not drastically improved, and in general, delicate and well controlled environment is needed to sustain the form of M-DNA.

Besides the double-helix, certain sequences can also adopt three- or four-stranded conformations [70, 71]. Four-stranded conformation is especially stable for guanine-rich sequences in the presence of monovalent and/or divalent metal cations [71–73]. These long structures, named G-wires, are comprised of stacked tetrads arising from the planar association of four guanines by Hoogsteen bonding [74, 75]. G-wire is a promising candidate for an electrical conductor since it is sturdier compared to dsDNA and made solely of guanine, characterized by the lowest ionization potential among the DNA bases, thus likely to enable more efficient charge migration along DNA. There already exists clear experimental evidence of an electrostatic polarizability of the G-wires indicating possible electrical conductivity [67].

In addition to its presumably better conductivity and improved mechanical properties, the G-wire still possesses almost the same self-assembly properties as dsDNA, and thus, can be equally functionalized or modified [76]. Functionalization of G-wires with gold and silver nanoparticles to form stable complexes has already been demonstrated [77, 78], and moreover, plenty of ideas about nanoscale molecular machines utilizing G-rich strands and their conformation changes have been suggested [79–83].

2.3 DNA-Based Electronic Biodevices

Self-assembled DNA-based devices can also be exploited as electronic bio-sensors, e.g. for recognizing DNA and certain base sequences [35, 84–87]. The working principle of these sensors can be based on electrochemical detection [88], direct electrical signal [89, 90] or for example on a DNA field effect transistor (DNA-FET) [91], where the gate is made of ssDNA molecules acting as surface receptors for investigated molecules. The latter one is based on the change of the charge distribution in the vicinity of the gate when a target molecule hybridizes with the receptors, and thus the current between the drain and the source will be modulated.

Solid-state nanopores are often used for electronic detection of various types of molecules [92], but for some particular applications the pore size, properties and functionality of the opening should be accurately tuned. This can be achieved by incorporation of DNA structures into the pores, mimicking the idea of protein pores in solid-state openings [93]. Very recent examples show that the 3D DNA origami structures can serve as plugs [94] or gatekeepers [95] for the lithographically fabricated pores. These hybrid pores can be precisely controlled in size and shape, and are easily functionalized. In addition, the extension of these methods demonstrated origami pores attached even to lipid membranes [96]. It is highly possible that combination of DNA transistors with nanopore techniques will lead to a realization of devices allowing cheap DNA sequencing in the near future.

3 Placement of DNA Structures on a Chip

In order to reliably determine the conductance of DNA structures or make any use of them in molecular electronics, they have to be integrated to other circuitry in a controllable way. Naturally, there exist numerous possible ways to achieve this, and only some of the most studied and sufficient methods are discussed in the following sections.

3.1 Anchoring DNA Structures on Patterned or Chemically Modified Surfaces

There exist many readily available chemical methods for positioning DNA-templates on the chip. One impressive example is to anchor DNA structures on lithographically fabricated wells according to their specific shape. Kershner et al. proposed and demonstrated how to attach triangular origamis to the origami-shaped binding sites etched in silicon oxide and diamond-like carbon

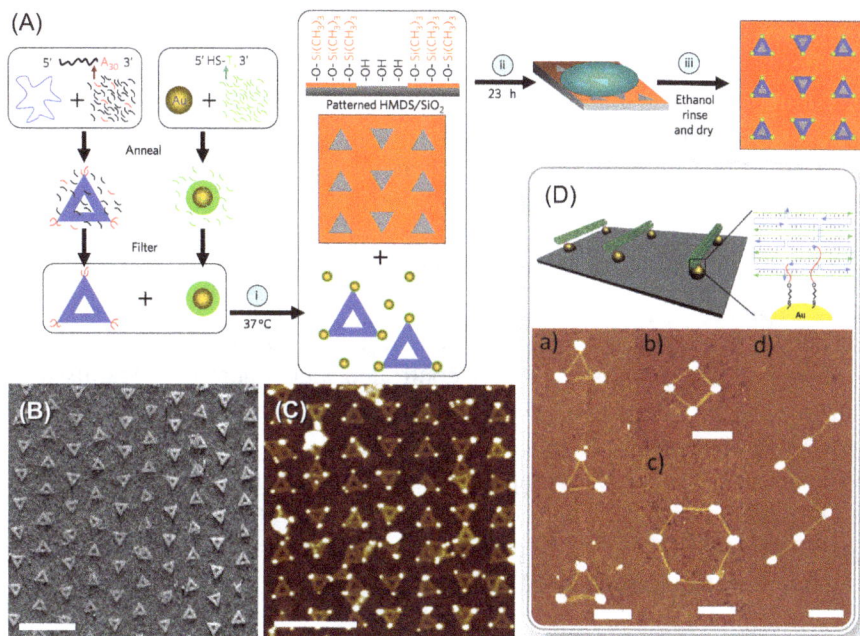

Figure 1 (A) Process for fabricating triangular DNA origamis with extra poly-adenine A_{30} strands extending from the corners, conjugating them with a poly-thymide functionalized gold nanoparticles (AuNP), and assembling two-dimensional nanoparticle arrays by utilizing triangular binding sites of clean oxide in a HMDS film patterned by electron-beam lithography [98]. The schematic illustrates the three key steps: (i) high-yield origami and AuNP binding, (ii) controlled DNA origami adsorption and (iii) ethanol treatment for drying and salt removal. (B) AFM image of triangular DNA origamis attached to the specific binding sites with a preferred orientation. The binding sites have sides of 110 nm and alternate between columns pointed up and columns pointed down. (C) AFM image of origamis bound with poly-T-coated AuNPs and adsorbed to a similar substrate with binding site side length of 100 nm. Scale bars in (B) and (C) are 500 nm. (D) Schematic drawing of lithographically fabricated gold islands connected by DNA origami tubes on the substrate. Thiolated DNA strands (red) are extended from each end of the DNA origami tube thus aligning the tubes along the gold islands. The position of the thiolated groups is designed so that the tube can only connect the gold islands, if its length matches the distance between the islands [101]. Below are AFM images of various structures formed by connecting gold islands with DNA origami tubes. All scale bars are 300 nm. (A)–(C) adapted from [98] by permission from MacMillan Publishers Ltd., © 2009 Nature Publishing Group. (D) Adapted from [101] with permission, © 2010 American Chemical Society.

substrates [97]. The followed extension of this method allowed one to organize gold nanoparticles on a chip with nanometer-scale resolution as shown in Figures 1A–C [98]. Besides the small feature size, this technique also provides high yield and enables a large scale assembly, thus being a candidate for commercial fabrication method of nanoelectronic devices in the future. Nanoparticles can also be assembled within the confined spaces with the help of patterned DNA strands in order to form surface-driven superlattices [99].

Other techniques for placement of DNA structures are often based on chemical attachment. Gerdon et al. used lithographically produced and 11-mercaptoundecanoic acid (MUA) modified areas on a chip for immobilizing origamis specifically, and furthermore positioning gold nanoparticles to the selected locations on top of the anchored origami [100]. There also exist other examples of chemicals suitable for controlled attachment of origamis on a substrate: hexamethyldisilazane (HMDS) prevents the attachment to certain areas [101, 102] and hydrogen silsesquioxane (HSQ) immobilizes origamis to the surfaces [103]. Lithographic methods (conventional or nanoimprint) can be utilized in patterning chemically selective binding areas on a large scale [101, 102, 104] or in fabrication of arrays of binding points attaching origamis selectively by the size as presented in Figure 1D [101, 105].

3.2 Trapping with Electric Fields

One of the most useful methods to direct and trap objects in solution is dielectrophoresis (DEP). It offers more dynamics and extra control on the trapping, if compared to the chemical or lithographical methods. DEP means a translational motion of a polarizable particle within an inhomogeneous electric field [106, 107]. The DEP force is proportional to the gradient of the square of the electric field, and the direction of the force depends on the polarizability of the particle compared to the surrounding medium. If the gradient and the difference in polarizabilities of the object and the medium are large enough, DEP can be utilized in manipulating materials even in nanoscale. Although the Brownian motion poses challenges in capturing of small objects from solution, various micro- and nanoscale objects have been successfully directed and trapped by DEP, and the method has been applied in variety of fields. DEP has largely been used as an active and non-destructive manipulation method for trapping cells, viruses, proteins and beads [108–110] as well as components directly exploitable for molecular electronics such as carbon nanotubes [111], nanoparticles [112, 113] and quantum dots [114] (see [115] for more complete review).

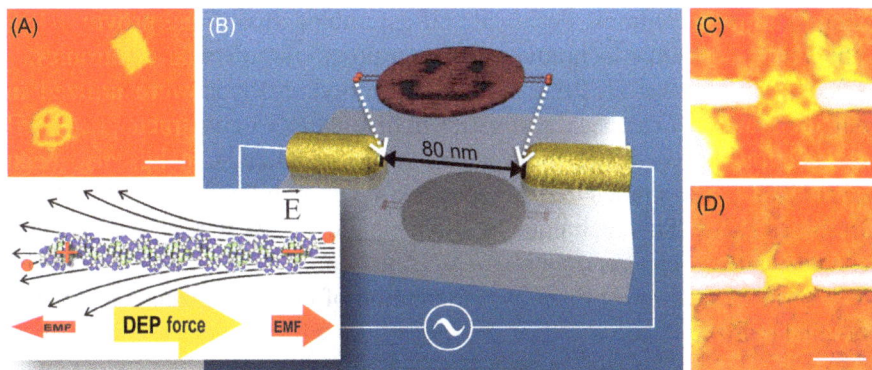

Figure 2 Trapping DNA origami with dielectrophoresis [123]. (A) AFM image of origami structures used for DEP trapping. The image is taken on a mica surface using tapping mode AFM in liquid. (B) Schematic view of the origami trapping using lithographically fabricated gold nanoelectrodes. The inset illustrates the electromagnetic forces (EMF) acting on a DNA and the principle of positive DEP. The DNA structures are also straightened during the DEP trapping. AFM image of a single smiley (C) and a rectangular origami (D) trapped with DEP (on SiO_2 surface, tapping mode AFM in air). All the scale bars are 100 nm. Adapted with permission from [123], © 2008 Wiley.

In case of DNA, even short fragments can be efficiently trapped, since the DNA is surrounded by a highly polarizable counter-ion cloud in a solution [116, 117], making the trapping more efficient [118, 119]. There exist a huge number of examples of trapping of DNA molecules by dielectrophoresis – starting from Masao Washizu's work in 1990s [120] – which are summarized in [119, 121, 122]. To average the electrophoretic forces due to the negative charge of DNA in aqueous buffer to zero, an AC voltage is usually utilized for trapping. By exploiting DEP-immobilization between nanoscale electrodes, DNA molecules can also be integrated and connected into the other circuitry.

The same methods can be equally applied to DNA-based structures. Kuzyk et al. showed that individual DNA origamis (smileys and rectangles [9]) can be trapped and immobilized to a silicon oxide chip in a controllable way by alternative current -DEP [123] as illustrated in Figure 2. The $\sim$12 MHz, $\sim$1 V_{pp} AC-voltage was applied to lithographically fabricated narrow fingertip-type gold nanoelectrodes, which yielded high enough gradient in the electrode gap for trapping the origamis. Origamis were thiol-modified in or-der to ensure a proper attachment to the electrodes via covalent sulphur-gold bonds. Trapping of the DNA origamis on a chip was the first reported DEP-manipulation of complex, designed self-assembled structures, proving DEP to be a truly adaptable technique for the purposes of molecular electronics.

Later on, similar DEP-based immobilization methods were utilized in characterization of electrical properties of distinct DNA constructs [124,125] (see Section 4). By using multiple electrode geometries, it has also been shown that DNA strands can be specifically immobilized only to selected electrodes [126]. This technique would enable complex wiring and bridging schemes of electrodes for creation of DNA networks [127], and prospectively, more sophisticated attachment and orientation of DNA templates on the chip could be realized.

4 Electrical Properties of DNA-Based Structures

As discussed above, it seems that dsDNA is a poor conductor. However, these results do not directly reveal the electrical properties of self-assembled dsDNA-based motifs. In Rothemund's original article [9], the topology of the adjacent dsDNA-like components in DNA origami was not yet resolved. Only very recently, it was shown that DNA can actually have previously unobserved and unnatural topologies in the scaffolded densely packed structures [22]. Therefore, DNA motifs can also support slightly different base stacking than the natural dsDNA, and thus, also their conductivity properties can be distinct. Moreover, dsDNA segments within the core of the DNA constructs can be structurally very well shielded from the external environment by the neighboring strands, and that could also prevent excess dehydration and helical conformation from collapsing. Thus, the influence of the environment (water, ions) to strand conformation might significantly vary between spatially distinct segments of the object. Apart from this, the role of the crossovers (strand exchanges between the neighboring helices in DNA objects) in total conductance of the structure is also unclear.

Since the DNA structures have already shown to possess a huge potential as templates in the nanoscale patterning, their electrical properties should also be fully understood for realization of the prospective nanoelectronic applications. The following sections discuss in more detail the conductivity properties of DNA motifs based purely on dsDNA, as well as the fabrication and electrical characterization of DNA-templated CNT-transistors as an example of successful utilization of the motifs in organizing materials.

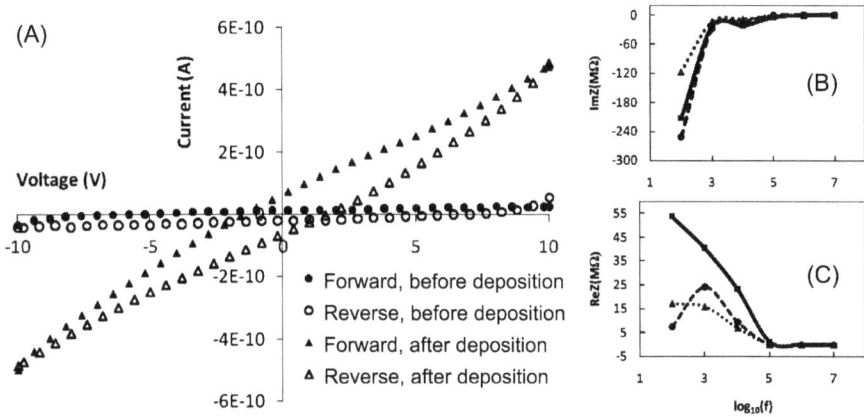

Figure 3 (A) Current-voltage characteristics before and after the deposition of the triangu-lar DNA origami between the electrodes on the chip. Resistance before the deposition is ∼400 GΩ and after ∼20 GΩ. Imaginary (B) and real (C) part of the measured impedancies of an empty chip (dashed line), chip with DNA origami deposited (dotted line), and impedance of pure DNA origami (solid line) as a function of frequency calculated assuming parallel connection of the chip and the origami [128]. Adapted with permission from [128], © 2009 American Institute of Physics.

4.1 dsDNA-Based Motifs

The very first measurement of the electrical properties of DNA origami was reported in 2009, when Bobadilla et al. determined the current through vari-ety of triangular origamis (not controlled number of structures) placed in the gap between voltage biased electrodes at ambient conditions [128]. The obtained DC-resistance of the parallel origamis was about 20 GΩ, as shown in Figure 3A. The group also measured the complex impedance within a wide range of frequencies. At low frequencies the impedance was high (similar to the DC-resistance) but was reduced with the increasing frequency reaching a significantly lower flat value at 100 kHz. Simultaneously the impedance turned from capacitive to resistive (see Figures 3B and C), suggesting that the DNA structures could be conductive in the high frequency region. It was also concluded that the conductivity of the DNA structures could not be determined alone or separately, since the attachment of origamis affect the conductance of the empty electrode arrangement as well.

The same group also measured a temperature dependence of the DC current-voltage characteristics of similar samples [129], revealing fully insu-lating behaviour below 240 K (resistance similar to the empty electrodes) and exponential dependency of the resistance with respect to the inverse of temperature close to the room temperature. In addition, the thermionic emission, as well as the hopping conduction models were fitted to the results yielding voltage dependent activation barriers ~0.7 and ~1.1 eV for low and high voltage regimes, respectively. Furthermore, the hopping was assigned as the main conductance mechanism close to the room temperature.

During the same year, Linko et al. characterized conductance mechanisms of single rectangular origamis [9] by utilizing DEP-immobilization described above [124]. A thiol-linker -modified and ligated [130] origami was immobilized between the nanoelectrodes (verified by AFM imaging), and both DC and AC characteristics were investigated at different relative humidity (RH) levels, since previous results suggested that distinct humidity conditions can have a huge influence to the conductivity of DNA [57–60]. At low RH levels the DNA origami was insulating with the resistance of the order of $T\Omega$ (similar resistance observed also for dry dsDNA [57]). At RH = 90%, DC-sweeping from −0.3 to 0.3 V produced non-linear current-voltage (*I–V*) curves with a resistance of 10 GΩ between −0.2 and 0.2 V, and about 2 GΩ outside this region as shown in Figure 4A. In comparison, the control sample (also underwent DEP, but without any DNA in the trapping buffer) yielded a linear *I–V* curve with a typical resistance value of 10–30 GΩ. Similar non-linear *I–V* characteristics have previously been reported also for dsDNA molecules, e.g. in [131, 132]. The high impedance of the DC measurement could be explained by the used hexanethiol-ssDNA linkers, as the resistance of hexanethiol has been reported to be from 10 MΩ to 1 GΩ [133] and, moreover, it has also been observed that a ssDNA molecule is a poor conductor [59]. The DC conductance was also determined as a function of RH, and the results indicated the conductance to be mostly ionic with a major contribution from the ionized water molecules [60, 134]. However, this observation could also be due to the highly resistive linkers conducting only via some water-assisted mechanism(s).

In addition, complex impedances of the same samples were measured at RH = 90% while varying the frequency of the AC bias voltage between 0.01 Hz and 100 kHz (Figures 4B and C). The AC impedance spectroscopy (AC-IS) results were modeled by equivalent circuits (Figures 4D and E) allowing one to identify distinct contributions to the total conductance [135]. The model was consistent with the DC data and also revealed that the resistance of the DNA origami or the DNA-assisted resistance in the gap region was about 70 MΩ. The measurement also showed that the conductance of a DNA origami in high humidity conditions is a combination of ohmic and ionic contributions and that the high impedance contacts can hide the actual

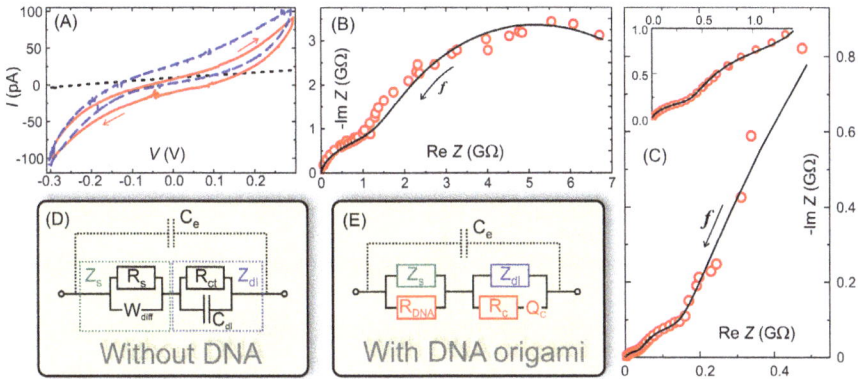

Figure 4 Impedance spectroscopy of a single rectangular DNA origami [124]. (A) Current-voltage characteristics of a control sample, underwent full DEP procedure except without DNA in the trapping buffer (black dotted line), and two samples containing single rectangular origamis between the electrodes (red solid line and blue dashed line). The hysteresis is due to the high humidity RH $\approx 90\%$. The direction of the voltage sweep is indicated by arrows. Cole-Cole plots of (B) control sample and (C) origami sample measured by impedance spectroscopy (red circles). The black lines are fittings of the equivalent circuits shown in (D) and (E). The arrows indicate the direction of increasing frequency. The inset in (C) is a blow out of the data near the origin. (D) Equivalent circuit for the control sample. C_e is the geometrical capacitance of the electrodes, and parallel combination of R_s, representing the small leakage current through the "electrolyte" (humid SiO_2 surface), and W_{diff}, describing the diffusion of the ions on the surface, forms the series impedance Z_s (green). In series with this, there is the double-layer capacitance, C_{dl}, and R_{ct} representing the current through it by redox reactions or tunneling. Together they form a double-layer impedance Z_{dl} (blue). (E) Equivalent circuit for the origami sample includes the full control sample circuit, and additional components due to origami: resistance of the origami R_{DNA} in parallel to Z_s and contact of the origami to the electrode described by combination of resistance R_c and a constant phase element Q_c in parallel to Z_{dl}. This combination of R_c and Q_c is a common way to describe many parallel connections with different time constants [135]. Adapted with permission from [124], © 2009 Wiley.

conductance of an investigated object in the DC-measurements. This was the first demonstration of a fully detailed equivalent circuit modelling for DNA or single DNA structures. The model also reasonably agreed with the results by Bobadilla et al. [128] and Bellido et al. [129]: DC resistances were similar and dependent on water, and in both studies at room temperature the impedance was reduced significantly in the higher frequency regime again indicating water-assisted conductance. The observed results were also comparable to other works for dsDNA molecules, e.g. [136, 137].

Since the conductance of DNA is known to drop with increasing length, similar AC-IS studies and modeling were performed for a shorter and smaller TX-tile-based molecular template [125]. This time hexanethiol-linkers were directly attached to the structure without low-conductance ssDNA linkers. However, the results showed that the impedance over the whole frequency range was higher than in the case of rectangular origamis. This observation might imply that the conductivity of DNA structures scales with the volume of the construct indicating water-induced or water-assisted conductivity along the DNA helices (polarised water molecules sheathing the DNA) to be the most probable charge transfer mechanism [59–61, 134].

In summary, the conductivity of the above-mentioned DNA structures (triangular and rectangular origamis and TX-tile-based objects) was found to be small suggesting that almost any kinds of devices for molecular electronics could be built on DNA without having to take the scaffold into account. The results indicate that the direct electronic conductivity via base pairs could be considered negligible in 2D DNA templates at least in the utilized setups and environments. This can simply be due to the non-optimal base stacking of the base pairs in a DNA structure [22, 40]. However, the conductance of 3D DNA structures still remains an open question. So far there is only one reported result of DNA-mediated CT in a 3D structure [138], but again, more results are expected soon.

4.2 DNA-Structures as Templates: CNT-Transistors

Despite the poor conductance of the reported sole DNA structures, various DNA templates can serve as nanoscale circuitboards for other electronic components, such as carbon nanotubes (CNTs), which are known to have suitable properties for molecular electronics and sensing [139]. Yet, it has been shown that, in addition to exploiting the DNA motifs in directing the CNTs to form molecular scale transistors [140–142], DNA can be also utilized in separation of different types of CNTs [143–145]. By combining these properties one can form an efficient tool-set for fabrication of CNT-transistors.

About a decade ago Keren et al. were the first ones to demonstrate the fabrication of a DNA-templated CNT-FET [146]. For a template they utilized a single dsDNA molecule, since more robust DNA motifs like origamis were not yet invented. The starting point in the fabrication was ∼200 nm long RecA modified ssDNA, which was hybridized to a selected point of much longer dsDNA template via homologous recombination [147]. Subsequently, biotinylated antibodies were attached to the RecA proteins, thus forming a

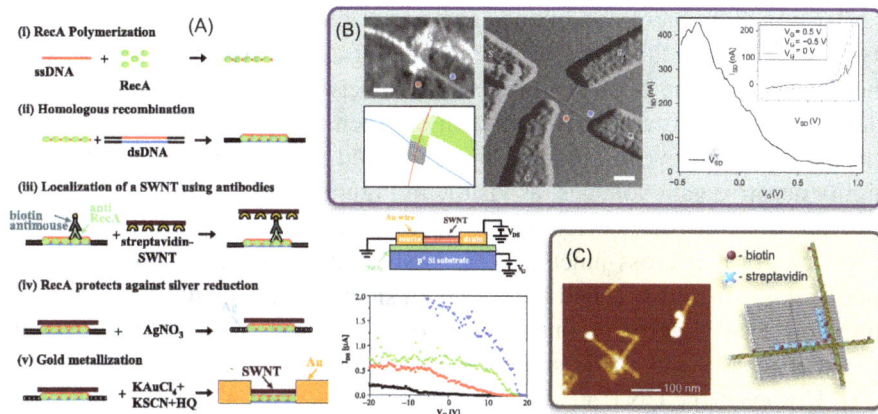

Figure 5 (A) Step by step assembly of a DNA-templated CNT-FET (i)–(v) [146]. Schematic representation of the electrical measurement circuit, and measured electrical characteristics: drain-source current (I_{DS}) versus gate voltage (V_G) for different values of drain-source bias: $V_{DS} = 0.5$ V (black), 1 V (red), 1.5 V (green), 2 V (blue). (B) Electrical characterization of a self-assembled CNT cross-junction [23]. *Left*: AFM image of a CNT cross-junction, and its schematic presentation on the ribbon (dark green) modified origami (grey). *Middle*: AFM amplitude image of the cross-junction with electron-beam patterned electrodes. The DNA template is no longer visible. Scale bars are 100 nm. *Right*: Source-drain current (I_{SD}) versus CNT gate voltage (V_G) for a source-drain bias of 0.85 V. Inset shows the source-drain *I–V*s for different gate voltages. (C) AFM image and schematics of a CNT assembly on DNA origami template using STV-biotin interaction [148]. CNTs wrapped with biotinmodified ssDNA were immobilized via STV on the origami templates with a certain pattern of biotin modifications. Adapted with permission: (A) from [146], © 2003 American Association for the Advancement of Science, (B) from [23], © 2009 Nature Publishing Group, and (C) from [148], © 2011 Wiley.

chain of biotin binding sites. A streptavidin (STV)-coated CNT was then attached to these sites, and finally the parts of the dsDNA without RecA modification were chemically metallized yielding gold electrodes attached to the both ends of the CNT. The whole fabrication process is illustrated in Figure 5A.

After fabrication, the DNA-templated CNT-FETs were mapped by AFM imaging and the gold electrodes were contacted by electron beam lithography, which allowed direct measurement of the devices, as illustrated in Figure 5A. The silicon substrate performed as a common gate electrode for all the devices on the same chip. In the same figure the measured current as a function of the gate voltage is represented for different bias (drain-source) voltages. The curves show clearly *p*-type FET behavior, typical for most

CNT-FETs at ambient conditions. This result clearly proves the applicability of DNA-based fabrication process via self-assembly. Only drawback in the process was the high contact resistance between the CNT and the gold electrodes, which is due to a mismatch between the gold work function and the CNT energy levels, inducing the undesired saturation of the current at highly negative gate voltages, as visible in Figure 5A.

Later on, the DNA origami has been exploited in organization of two CNTs to form a cross-junction [23, 148]. Maune et al. used nucleotide-modified CNTs and attached them to both sides of a rectangular origami [9] via short DNA overhangs attained as perpendicular rows on the origami by extensions of the staple strands pointing out from the origami template [23]. However, the CNT attachment was successful only after extending the origami to a ribbon as shown in Figure 5B.

After depositing on a SiO_2 covered silicon substrate, CNTs were connected by lithographically fabricated Au/Pt electrodes. Palladium (Pt) was chosen to minimize the contact resistance due to the work function, and furthermore, the exposed ends of the CNTs were chemically cleaned from DNA before depositing the metal. The applicability of the formed cross-junction as a transistor was demonstrated by characterizing its electrical properties using one CNT as a current channel of the FET and other as a gate. In total, six CNT-FETs were fabricated. Current-voltage characteristics measured from one of the FETs are shown as an example in the inset of the right panel of the figure 5B with different gate voltages. The main frame illustrates the observed gate voltage dependence revealing again the typical *p*-type behaviour. In this case the saturation at higher negative voltages was absent due to the smaller contact resistance.

Recently, Eskelinen et al. demonstrated an alternative and efficient method – based on biotin-STV interaction – for forming a similar CNT cross-junction on a DNA origami [148]. In that study, certain locations of the DNA origami were functionalized with biotin and followed by STV attachment. The formed streptavidin pattern again allowed biotinylated ssDNA-wrapped CNTs to be attached and aligned on the DNA template as shown in Figure 5C.

Apart from successful templating of CNT-transistors, DNA origamis have also been shown to be suitable platforms for creating plasmonic structures [28], for imaging and analyzing of single molecules [149–151], and for guiding chemical reactions [24], just to mention a few examples. Many more applications are foreseen.

5 Summary and Outlook

DNA has indeed proven to offer an ever increasing variety of possibilities to be utilized in fabrication of nanoscale structures. Its striking self-assembly capabilities has driven researchers to develop more and more novel ideas, driving the whole field through a major maturation; during the last two decades a development from the delicate tile-based systems to more robust origami-based methods has been witnessed. These approaches based on exploiting the exceptional self-assembly characteristics of DNA can serve as a toolbox for the next generation of device fabrication enabling the production of nanostructures made of materials relevant for electronics, optics and sensing. By combining the efficient and controllable DEP manipulation and chemical as well as geometrical placement methods of complex DNA constructs with the top-down techniques one can fabricate sophisticated and highly ordered circuits and functional devices truly in molecular scale.

In most cases the conductivity of the DNA scaffold could be considered negligible when used to assemble other molecular components. However, the conductivity of the DNA itself is still not fully untangled. Doping with metal ions, or on the other hand, the new forms of DNA, like G-wires; or even a recently discovered plasmon-initiated long range excitation transfer along a dsDNA [152], might open up opportunities for realizing novel DNA-based devices. Thus, the story of self-assembled DNA-based structures for molecular electronics is not yet fully written.

Acknowledgements

Financial support from Academy of Finland (Project Nos. 218182 and 263262) is greatly acknowledged. V.L. thanks the Emil Aaltonen Foundation. J.J.T. thanks NewIndigo ERA-NET NPP2 (AQUATEST, INDIGO-DST1-012) and EU's COST action MP0802 for enabling fruitful collaborations.

References

[1] N. C. Seeman, J. Theor. Biol., 99, 237–247 (1982).
[2] N. R. Kallenbach, R.-I. Ma, N. C. Seeman, Nature, 305, 829–831 (1983).
[3] E. Winfree, F. Liu, L. A. Wenzler, N. C. Seeman, Nature, 394, 539–544 (1998).
[4] H. Yan, S. H. Park, G. Finkelstein, J. H. Reif, T. H. LaBean, Science, 301, 1882–1884 (2003).
[5] P. W. K. Rothemund, A. Ekani-Nkodo, N. Papadakis, A. Kumar, D. K. Fygenson, E. Winfree, J. Am. Chem. Soc., 126, 16344–16352 (2004).

[6] P. W. K. Rothemund, N. Papadakis, E. Winfree, PLoS Biol., 2, e424 (2004).

[7] H. Yan, T. H. LaBean, L. Feng, J. H. Reif, Proc. Natl. Acad. Sci. U.S.A., 100, 8103–8108 (2003).

[8] W. M. Shih, J. D. Quispe, G. F. Joyce, Nature, 427, 618–621 (2004).

[9] P. W. K. Rothemund, Nature, 440, 297–302 (2006).

[10] S. M. Douglas, H. Dietz, T. Liedl, B. Högberg, F. Graf, W. M. Shih, Nature, 459, 414–418 (2009).

[11] T. Liedl, B. Högberg, J. Tytell, D. E. Ingber, W. M. Shih, Nat. Nanotechnol., 5, 520–524 (2010).

[12] H. Dietz, S. M. Douglas, W. M. Shih, Science, 325, 725–730 (2009).

[13] D. Han, S. Pal, J. Nangreave, Z. Deng, Y. Liu, H. Yan, Science, 332, 342–346 (2011).

[14] S. M. Douglas, A. H. Marblestone, S. Teerapittayanon, A. Vasquez, G. M. Church, W. M. Shih, Nucleic Acid Res., 37, 5001–5006 (2009).

[15] C. E. Castro, F. Kilchherr, D.-N. Kim, E. Lin Shiao, T. Wauer, P. Wortmann, H. Dietz, Nat. Meth., 8, 221–229 (2011).

[16] D.-N. Kim, F. Kilchherr, H. Dietz, M. Bathe, Nucleic Acid Res., 40, 2862–2868 (2012).

[17] A. V. Pinheiro, D. Han, W. M. Shih, H. Yan, Nat. Nanotechnol., 6, 763–772 (2011).

[18] T. Tørring, N. V. Voigt, J. Nangreave, H. Yan, K. V. Gothelf, Chem. Soc. Rev., 40, 5636–5646 (2011).

[19] J.-P. J. Sobczak, T. G. Martin, T. Gerling, H. Dietz, Science, 338, 1458–1461 (2012).

[20] B. Wei, M. Dai, P. Yin, Nature, 485, 623–626 (2012).

[21] Y. Ke, L. L. Ong, W. M. Shih, P. Yin, Science, 338, 1177–1183 (2012).

[22] X.-c. Bai, T. G. Martin, S. H. W. Scheres, H. Dietz, Proc. Natl. Acad. Sci. U.S.A., 109, 20012–20017 (2012).

[23] H. T. Maune, S. Han, R. D. Barish, M. Bockrath, W. A. Goddard III, P. W. K. Rothemund, E. Winfree, Nat. Nanotechnol., 5, 61–66 (2010).

[24] N. V. Voigt, T. Tørring, A. Rotaru, M. F. Jacobsen, J. B. Ravnsbæk, R. Subramani, W. Mamdouh, J. Kjems, A. Mokhir, F. Besenbacher, K. V. Gothelf, Nat. Nanotechnol., 5, 200–203 (2010).

[25] A. Kuzyk, K. T. Laitinen, P. Törmä, Nanotechnology, 20, 235305 (2009).

[26] R. Schreiber, S. Kempter, S. Holler, V. Schüller, D. Schiffels, S. S. Simmel, P. C. Nickels, T. Liedl, Small, 7, 1795–1799 (2011).

[27] M. R. Jones, K. D. Osberg, R. J. Macfarlane, M. R. Langille, C. A. Mirkin, Chem. Rev., 111, 3736–3827 (2011).

[28] A. Kuzyk, R. Schreiber, Z. Fan, G. Pardatscher, E.-M. Roller, A. Högele, F. C. Simmel, A. O. Govorov, T. Liedl, Nature, 483, 311–314 (2012).

[29] S. J. Tan, M. J. Campolongo, D. Luo, W. Cheng, Nat. Nanotechnol., 6, 268–276 (2011).

[30] R. Wang, C. Nuckolls, S. J. Wind, Angew. Chem. Int. Ed., 51, 11325–11327 (2012).

[31] E. Braun, Y. Eichen, U. Sivan, G. Ben-Yoseph, Nature, 391, 775–778 (1998).

[32] J. Liu, Y. Geng, E. Pound, S. Gyawali, J. R. Ashton, J. Hickey, A. T. Woolley, J. N. Harb, ACS Nano, 5, 2240–2247 (2011).

[33] Y. Geng, J. Liu, E. Pound, S. Gyawali, J. N. Harb, A. T. Woolley, J. Mater. Chem., 21, 12126–12131 (2011).

[34] International Technology Roadmap for Semiconductors 2011 Edition, http://www.itrs.net/Links/2011ITRS/Home2011.htm

[35] A. Csáki, G. Maubach, D. Born, J. Reichert, W. Fritzsche, Single Mol., 3, 275–280 (2002).

[36] K. Galatsis, K. L. Wang, M. Ozkan, C. S. Ozkan, Y. Huang, J. P. Chang, H. G. Monbouquette, Y. Chen, P. Nealey, Y. Botros, Adv. Mater., 22, 769–778 (2010).

[37] K. Sakakibara, J. P. Hill, K. Ariga, Small, 7, 1288–1308 (2011).

[38] J. Retèl, B. Hoebee, J. E. Braun, J. T. Lutgerink, E. van den Akker, A. H. Wanamarta, H. Joenje, M. V. Lafleur, Mutat. Res., 299, 165–182 (1993).

[39] C. Dekker, M. Ratner, Phys. World, 14, 29–33 (2001).

[40] J. D. Slinker, N. B. Muren, S. E. Renfrew, J. K. Barton, Nat. Chem., 3, 228–233 (2011).

[41] D. D. Eley, D. I. Spivey, Trans. Faraday Soc., 58, 411–415 (1962).

[42] C. J. Murphy, M. R. Arkin, Y. Jenkins, N. D. Ghatlia, N. J. Turro, J. K. Barton, Science, 262, 1025–1029 (1993).

[43] T. J. Meade, J. F. Kayyem, Angew. Chem. Int. Ed., 34, 352–354 (1995).

[44] F. Lewis, T. Wu, Y. Zhang, R. Letsinger, S. Greenfield, M. Wasielewski, Science, 277, 673–676 (1997).

[45] E. M. Boon, J. K. Barton, Curr. Opin. Struct. Biol., 12, 320–329 (2002).

[46] H.-W. Fink, C. Schönenberger, Nature, 398, 407–410 (1999).

[47] D. Porath, A. Bezryadin, S. de Vries, C. Dekker, Nature, 403, 635–638 (1999).

[48] A. Y. Kasumov, M. Koziak, S. Guéron, B. Reulet, V. T. Volkov, D. V. Klinov, H. Bouchiat, Science, 291, 280–282 (2001).

[49] X. Guo, A. A. Gorodetsky, J. Hone, J. K. Barton, C. Nuckolls, Nat. Nanotech., 3, 163–167 (2008).

[50] E. Shapir, H. Cohen , A, Calzolari , C. Cavazzoni, D. A. Ryndyk , G. Cuniberti, A. Kotlyar , R. Di Felice, D. Porath Nat. Mater., 7, 68–74 (2008).

[51] R. G. Endres, D. L. Cox, R. R. P. Singh, Rev. Mod. Phys., 76, 195–214 (2004).

[52] D. Porath, G. Cuniberti, R. Di Felice, Top Curr. Chem., 237, 183–228 (2004).

[53] M. Di Ventra, M. Zwolak, Encycl. of Nanosci. and Nanotechnol., 2, 475–493 (2004).

[54] K. W Hipps, Science, 294, 536–537 (2001).

[55] X. Guo, J. P. Small, J. E. Klare, Y. Wang, M. S. Purewal, I. W. Tam, B. Hee Hong, R. Caldwell, L. Huang, S. O'Brien, J. Yan, R. Breslow, S. J. Wind, J. Hone, P. Kim, C. Nuckolls, Science, 311, 356–358 (2006).

[56] J. M. Warman, M. P. deHaas, A Rupprecht, Chem. Phys. Lett., 249, 319–322 (1996).

[57] S. Tuukkanen, A. Kuzyk, J. J. Toppari, V. P. Hytönen, T. Ihalainen, P. Törmä, Appl. Phys. Lett., 87, 183102 (2005).

[58] J. Berashevich, T. Chakraborty, J. Phys. Chem. B, 112, 14083–14089 (2008).

[59] T. Kleine-Ostmann, C. Jördens, K. Baaske, T. Weimann, M. H. de Angelis, M. Koch, Appl. Phys. Lett., 88, 102102 (2006).

[60] C. Yamahata, D. Collard, T. Takekawa, M. Kumemura, G. Hashiguchi, H. Fujita, Biophys. J., 94, 63–70 (2008).

[61] M. Briman, N. P. Armitage, E. Helgren, G. Grüner, Nano Lett., 4, 733–736 (2004).

[62] A. Y. Kasumov, D. V. Klinov, P.-E. Roche, S. Guéron, H. Bouchiat, Appl. Phys. Lett., 84, 1007 (2004).

[63] Y.-T. Long, C.-Z. Li, H.-B. Kraatz, J. S. Lee, Biophys. J., 84, 3218–3225 (2003).

[64] S. Liu, G. H. Clever, Y. Takezawa, M. Kaneko, K. Tanaka, X. Guo, M. Shionoya, Angew. Chem. Int. Ed., 50, 8886–8890 (2011).

[65] E. Shapir, G. Brancolini, T. Molotsky, A. B. Kotlyar, R. Di Felice, D. Porath, Adv. Mater., 23, 4290–4294 (2011).

[66] A. B. Kotlyar, N. Borovok, T. Molotsky, H. Cohen, E. Shapir, D. Porath, Adv. Mater., 17, 1901–1905 (2005).

[67] H. Cohen, T. Shapir, N. Borovok, T. Molotsky, R. Di Felice, A. B. Kotlyar, D. Porath, Nano Lett., 7, 981–986 (2007).

[68] E. Shapir, L. Sagiv, T. Molotsky, A. B. Kotlyar, R. Di Felice, D. Porath, J. Phys. Chem. C, 114, 22079–22084 (2010).

[69] P. B. Woiczikowski, T. Kubar, R. Gutiérrez, G. Cuniberti, M. Elstner, J. Chem. Phys., 133, 035103 (2010).

[70] V. N. Soyfer, V. N. Potaman, in Triple-Helical Nucleic Acids (Springer, New York, 1995).

[71] G. N. Parkinson, M. P. Lee, S. Neidle, Nature, 417, 876–880 (2002).

[72] W. J. Qin, L. Y. Yung, Nucleic Acids Res., 35, e111 (2007).

[73] N. Borovok, N. Iram, D. Zikich, J. Ghabboun, G.I. Livshits, D. Porath, A. B. Kotlyar, Nucleic Acids Res., 36, 5050–5060 (2008).

[74] J. R. Williamson, M. K. Raghuraman, T. R. Cech, Cell, 59, 871–880 (1989).

[75] J. T. Davis, Angew. Chem., Int. Ed. Engl., 43, 668–698 (2004).

[76] S. Lyonnais, O. Piétrement, A. Chepelianski, S. Guéron, L. Lacroix, E. Le Cam, J.-L. Mergny, Nucl. Acids Symp. Ser., 52, 689–690 (2008).

[77] I. Lubitz, A. B. Kotlyar, Bioconjugate Chem., 22, 482–487 (2011).

[78] C. Leiterer, A. Csáki, W. Fritzsche, Methods and Protocols, Series: Methods in Molecular Biology, 749, 141–150. Eds: Giampaolo Zuccheri and Bruno Samorì (Humana Press, Springer, 2011).

[79] P. Alberti, J.-L. Mergny, Proc. Natl. Acad. Sci., 100, 1569–1573 (2003).

[80] R. P. Fahlman, M. Hsing, C. Sporer-Tuhten, D. Sen, Nano Lett., 3, 1073–1078 (2003).

[81] Z. S. Wu, C. R. Chen, G. L. Shen, R. Q. Yu, Biomater., 29, 2689–2696 (2008).

[82] B. Ge, Y. C. Huang, D. Sen, H.-Z. Yu, Angew. Chem. Int. Ed., 49, 9965–9967 (2010).

[83] X. Yang, D. Liu, P. Lu, Y. Zhangc, C. Yu, Analyst, 135, 2074–2078 (2010).

[84] W. Fritzsche, T. A. Taton, Nanotechnology, 14, R63 (2003).

[85] E. Souteyrand, J. P. Cloarec, J. R. Martin, C. Wilson, I. Lawrence, S. Mikkelsen, M. F. Lawrence, J. Phys. Chem. B, 101, 2980–2985 (1997).

[86] J. Fritz, E. B. Cooper, S. Gaudet, P. K. Sorger, S. R. Manalis, Proc. Natl. Acad. Sci. U.S.A., 99, 14142–14146 (2002).

[87] Z. Li, Y. Chen, X. Li, T. I. Kamins, K. Nauka, R. S. Williams, Nano Lett., 4, 245–247 (2004).

[88] E. E. Ferapontova, K. V Gothelf, Curr. Org. Chem., 15, 498–505 (2011).

[89] S.-J. Park, T. A. Taton, C. A. Mirkin, Science, 295, 1503–1506 (2002).

[90] R. Moeller and W. Fritzsche, IEE Proc.-Nanobiotechnol., 152, 47–51 (2005).

[91] N. Mohanty, V. Berry, Nano Lett., 8, 4469–4476 (2008).

[92] C. Dekker, Nat. Nanotechnol., 2, 209–215 (2007).

[93] A. R. Hall, A. Scott, D. Rotem, K. K. Mehta, H. Bayley, C. Dekker, Nat. Nanotechnol., 5, 874–877 (2010).

[94] N. A. W. Bell, C. R. Engst, M. Ablay, G. Divitini, C. Ducati, T. Liedl, U. F. Keyser, Nano Lett., 12, 512–517 (2012).

[95] R. Wei, T. G. Martin, U. Rant, H. Dietz, Angew. Chem. Int. Ed., 51, 4864–4867 (2012).

[96] M. Langecker, V. Arnaut, T. G. Martin, J. List, S. Renner, M. Mayer, H. Dietz, F. C. Simmel, Science, 338, 932–936 (2012).
[97] R. J. Kershner, L. D. Bozano, C. M. Micheel, A. M. Hung, A. R. Fornof, J. N. Cha, C. T. Rettner, M. Bersani, J. Frommer, P. W. K. Rothemund, G. M. Wallraff, Nat. Nanotechnol., 4, 557–561 (2009).
[98] A. M. Hung, C. M. Micheel, L. D. Bozano, L. W. Osterbur, G. M. Wallraff, J. N. Cha, Nat. Nanotechnol., 5, 121–126 (2010).
[99] H. Noh, A. M. Hung, J. N. Cha, Small, 7, 3021–3025 (2011).
[100] A. E. Gerdon, S. S. Oh, K. Hsieh, Y. Ke, H. Yan, H. T. Soh, Small, 5, 1942–1946 (2009).
[101] B. Ding, H. Wu, W. Xu, Z. Zhao, Y. Liu, H. Yu, H. Yan, Nano Lett., 10, 5065–5069 (2010).
[102] E. Penzo, R. Wang, M. Palma, S. J. Wind, J. Vac. Sci. Technol. B, 29, 06F205 (2011).
[103] F. A. Shah, K. N. Kim, M. Lieberman, G. H. Bernstein, J. Vac. Sci. Technol. B, 30, 011806 (2012).
[104] M. Palma, J. J. Abramson, A. A. Gorodetsky, E, Penzo, R. L. Gonzalez, Jr., M. P. Sheetz, C. Nuckolls, J. Hone, S. J. Wind, J. Am. Chem. Soc., 133, 7656–7659 (2011).
[105] A. C. Pearson, E. Pound, A. T. Woolley, M. R. Linford, J. N. Harb, R. C. Davis, Nano Lett., 11, 1981–1987 (2011).
[106] H. A. Pohl, J. Appl. Phys., 22, 869–871 (1951).
[107] H. A. Pohl, in Dielectrophoresis: The Behavior of Neutral Matter in Nonuniform Electric Fields (Cambridge Univesity Press, Cambridge, UK, 1978).
[108] M. P. Hughes, Nanotechnology, 11, 124–132 (2000).
[109] P. J. Burke, Encycl. of Nanosci. and Nanotechnol., 6, 623–641 (2004).
[110] L. Zheng, J. P. Brody, P. J. Burke, Biosens. Bioel., 20, 606–619 (2004).
[111] A. Vijayaraghavan, S. Blatt, D. Weissenberger, M. Oron-Carl, F. Hennrich, D. Gerthsen, H. Hahn, R. Krupke, Nano. Lett., 7, 1556–1560 (2007).
[112] R. Kretschmer, W. Fritzsche, Langmuir, 20, 11797-11801 (2004).
[113] S. Kumar, Y.-K. Seo, G.-H. Kim, Appl. Phys. Lett., 94, 53104 (2009).
[114] T. K. Hakala, V. Linko, A.-P. Eskelinen, J. J. Toppari, A. Kuzyk, P. Törmä, Small, 5, 2683–2686 (2009).
[115] A. Kuzyk, Electrophoresis, 32, 2307–2313 (2011).
[116] H. P. Schwan, G. Schwarz, J. Maczuk, H. Pauly, J. Phys. Chem., 66, 2626–2636 (1962).
[117] S. Suzuki, T. Yamanashi, S. Tazawa, O. Kurosawa, M. Washizu, IEEE Trans. Ind. Appl., 34, 75–83 (1998).
[118] S. Tuukkanen, A. Kuzyk, J. J. Toppari, H. Häkkinen, V. P. Hytönen, E. Niskanen, M. Rinkiö, P. Törmä, Nanotechnology, 18, 295204 (2007).
[119] R. Hölzel, IET Nanobiotechnol., 3, 28–45 (2009).
[120] M. Washizu, O. Kurosawa, IEEE Trans. Ind. Appl., 26, 1165–1172 (1990).
[121] R. Hölzel, F. F. Bier, IEE Proc.: Nanobiotechnol., 150, 47–53 (2003).
[122] A. Kuzyk, J. J. Toppari, P. Törmä, Methods and Protocols, Series: Methods in Molecular Biology, 749, 223–234. Eds: Giampaolo Zuccheri and Bruno Samorì (Humana Press, Springer, 2011).
[123] A. Kuzyk, B. Yurke, J. J. Toppari, V. Linko, P. Törmä, Small, 4, 447–450 (2008).
[124] V. Linko, S.-T. Paasonen, A. Kuzyk, P. Törmä, J. J. Toppari, Small, 5, 2382–2386 (2009).

[125] V. Linko, J. Leppiniemi, S.-T. Paasonen, V. P. Hytönen, J. J. Toppari, Nanotechnology, 22, 275610 (2011).

[126] V. Linko, J. Leppiniemi, B. Shen, E. Niskanen, V. P. Hytönen, J. J. Toppari, Nanoscale, 3, 3788–3792 (2011).

[127] Y. Eichen, E. Braun, U. Sivan, G. Ben-Yoseph, Acta Polym., 49, 663–670 (1998).

[128] A. D. Bobadilla, E. P. Bellido, N. L. Rangel, H. Zhong, M. L. Norton, A. Sinitskii, J. M. Seminario, J. Chem. Phys., 130, 171101 (2009).

[129] E. P. Bellido, A. D. Bobadilla, N. L. Rangel, H. Zhong, M. L. Norton, A. Sinitskii, J. M. Seminario, Nanotechnology, 20, 175102 (2009).

[130] P. O'Neill, P. W. K. Rothemund, A. Kumar, D. K. Fygenson, Nano Lett., 6, 1379–1383 (2006).

[131] H. Cohen, C. Nogues, R. Naaman, D. Porath, Proc. Natl. Acad. Sci, U.S.A., 102, 11589–11593 (2005).

[132] A. Rakitin, P. Aich, C. Papadopoulos, Y. Kobzar, A. S. Vedeneev, J. S. Lee, J. M. Xu, Phys. Rev. Lett., 86, 3670–3673 (2001).

[133] B. Xu, N. J. Tao, Science, 301, 1221–1223 (2003).

[134] D. H. Ha, H. Nham, K.-H. Yoo, H. So, H.-Y. Lee, T. Kawai, Chem. Phys. Lett., 355, 405–409 (2002).

[135] E. Barsoukov, J. R. MacDonald, in Impedance Spectroscopy: Theory, Experiment, and Applications, Second Edition (Wiley, Hoboken, New Jersey, 2005).

[136] J. Wang, Phys. Rev. B, 78, 245304 (2008).

[137] B. Xu, P. Zhang, X. Li, N. Tao, Nano Lett., 4, 1105–1108 (2004).

[138] N. Lu, H. Pei, Z. Ge, C. R. Simmons, H. Yan, C. Fan, J. Am. Chem. Soc., 134, 13148–13151 (2012).

[139] J. Kong, N. R. Franklin, C. Zhou, M. G. Chapline, S. Peng, K. Cho, H. Dai, Science, 287, 622–625 (2000).

[140] S. Lyonnais, C.-L. Chung, L. Goux-Capes, C. Escudé, O. Piétrement, S. Baconnais, E. Le Cam, J.-P. Bourgoin, A. Filoramo, Chem. Comm., 6, 683–685 (2009).

[141] P. F. Xu, H. Noh, J. H. Lee, J. N. Cha, Phys. Chem. Chem. Phys., 13, 10004–10008 (2011).

[142] A. D. Bobadilla, J. M. Seminario, J. Phys. Chem. C, 115, 3466–3474 (2011).

[143] M. Zheng, A. Jagota, E. D. Semke, B. A. Diner, R. S. Mclean, S. R. Lustig, R. E. Richardson, N. G. Tassi, Nat. Mater., 2, 338–342 (2003).

[144] X. Tu, S. Manohar, A. Jagota, M. Zheng, Nature, 460, 250–253 (2009).

[145] X. Tu, A. R. Hight Walker, C. Y. Khripin, M. Zheng, J. Am. Chem. Soc., 133, 12998–13001 (2011).

[146] K. Keren, R. S. Berman, E. Buchstab, U. Sivan, E. Braun, Science, 302, 1380–1382 (2003).

[147] K. Keren, M. Krueger, R. Gilad, G. Ben-Yoseph, U. Sivan, E. Braun, Science, 297, 72–75 (2002).

[148] A.-P. Eskelinen, A. Kuzyk, T. K. Kaltiaisenaho, M. Y. Timmermans, A. G. Nasibulin, E. I. Kauppinen, P. Törmä, Small, 7, 746–750 (2011).

[149] Y. Sannohe, M. Endo, Y. Katsuda, K. Hidaka, H, Sugiyama, J. Am. Chem. Soc., 132, 16311–16313 (2010).

[150] D. N. Selmi, R. J. Adamson, H. Attrill, A. D. Goddard, R. J. C. Gilbert, A. Watts, A. J. Turberfield, Nano Lett., 11, 657–660 (2011).

[151] M. J. Berardi, W. M. Shih, S. C. Harrison, J. J. Chou, Nature, 476, 109–113 (2011).
[152] J. Wirth, F. Garwe, G. Hähnel, A. Csáki, N. Jahr, O. Stranik, W. Paa, W. Fritzsche, Nano Lett., 11, 1505–1511 (2011).

Biographies

Veikko Linko received his M.Sc. (2007) and Ph.D. degrees (2011) in Physics from the University of Jyväskylä, Finland, under supervision of Dr. Jussi Toppari. During his studies he was a member of the Nanoelectronics group of Professor Päivi Törmä (2006–2007) and the Molecular Electronics & Plasmonics group of Dr. Jussi Toppari (2008–2011). He currently works as a postdoctoral researcher in Professor Hendrik Dietz's Laboratory for Biomolecular Nanotechnology at Technische Universität München in Germany. His research interests are self-assembled DNA nanostructures and their electrical properties, as well as hybrid protein-DNA -complexes.

J. Jussi Toppari (Academy Research Fellow) received his M.Sc. (1997) and Ph.D. degrees (2003) in Nanophysics from the University of Jyväskylä, Finland, under supervision of Professor Jukka Pekola. After that he worked as a senior assistant in a newly formed Nanoscience Center (NSC) of the

University of Jyväskylä. In 2008 he obtained the degree of adjunct professor (docent) and established his own independent research group in NSC. During 2011 he worked as a senior visiting researcher at the Institute of Photonic Technology, Jena, Germany. Research topics include, e.g., utilization of DNA self-assembled structures for fabrication of nanoscale electronic or plasmonic devices, as well as studies of strong coupling between surface plasmon polaritons and optically active molecules.

Molecular Combing of DNA: Methods and Applications

Zeinab Esmail Nazari and Leonid Gurevich*

Institute of Physics and Nanotechnology, Aalborg University, 9220 Aalborg, Denmark
Corresponding author: lg@nano.aau.dk

Received 11 December 2012; Accepted 15 January 2013

Abstract

First proposed in 1994, molecular combing of DNA is a technique that allows adsorption and alignment of DNA on the surface with no need for prior modification of the molecule. Since then, many variations of the original method have been devised and used in a wide range of applications from genomic studies to nanoelectronics. While molecular combing has been applied in a variety of DNA-related studies, no comprehensive review has been published on different combing methods proposed so far. In this review, the underlying mechanisms of molecular combing of DNA are described followed by discussion of the main methods in molecular combing as well as its major applications in nanotechnology.

Keywords: Molecular combing of DNA, DNA stretching, AFM.

1 Introduction

While most of our electronics devices currently rely on silicon technology, the Moor's law predicts that we are approaching its limits [1, 2], implying that we are still in search for new ideas. Nature is a perfect source of inspiration, in particular, from the perspective of complex and reliable machinery at nanoscale it employs. For instance, the multi-complex protein machinery

Journal of Self-Assembly and Molecular Electronics, Vol. 1, 125–148.

responsible for DNA replication (also known as replisome) is a nanomachine that polymerizes DNA at an amazing rate of up to 1000 nucleotides per second and makes less than one mistake per 10^9 nucleotide incorporations [3]. If we aim to achieve that level of complexity and precision in our future nanotechnology devices, probably the best way would be to bring our technology closer to that of nature. The very first step to achieve this would be to harvest the potential of the existing biological building blocks by incorporating them into our nanodevices and develop hybrid structures containing both organic and inorganic parts.

In this context, and for many reasons, DNA has been the center of attention in nanotechnology research, from genomic and biomedical studies [4, 5] to development of nanomachines and nanocircuits [6]. However, many investigations and manipulations are not possible on DNA molecule in its natural coiled structure. From this perspective, strategies that enable us to immobilize and straightening DNA on a solid substrate would be of great value since they open many possibilities for nanotechnology, for instance, fabrication of hybrid structures benefiting, on one hand, from unique features of self-assembly, recognition, and self-replication of DNA [7–9] and, on the other hand, from state-of-the-art fabrication procedures developed for inorganic materials. In order to achieve this aim, numerous methods have been introduced for aligning DNA molecules on solid surfaces using different approaches which involve stretching DNA using optical or magnetic tweezers [10], spin stretching [11, 12], or stretching inside nanofluidic channels [13]. Most of these techniques for stretching DNA require further modification of DNA extremities through biochemical reactions to anchor DNA to a functionalized substrate, e.g., via sticky ends or biotin functionalization [14]. An interesting approach for aligning DNA on solid surfaces is molecular combing of DNA, which is a reliable method for immobilization and stretching of DNA without the need for any prior modification to DNA extremities [15]. This technique was first introduced by Bensimon et al. in 1994 who also coined the name "molecular combing" [16]. It has been extensively used since then and many different protocols have been devised with the idea of combing: anchoring hairs at one end and using a force to comb them!

Combing can be defined as a procedure in which – under certain conditions of pH and ionic strength – DNA molecules are attached, at one or both ends, to the surface and subsequently aligned by the force applied at the air/water interface. Molecular combing offers a number of advantages over common stretching methods: as mentioned earlier, it allows stretching of DNA without the need to modify the ends of DNA, it is suitable for stretching a large number of DNA molecules as well as very long strands of DNA molecules, such as genomic DNA, and it is a facile and convenient procedure with reasonable reproducibility.

In the simplest form of this technique, a droplet of DNA solution is deposited on a hydrophobic surface. At proper values of pH and ionic strength, DNA molecules slightly unwind at the ends and expose the hydrophobic core (formed by stacked hydrophobic bases), which can attach to the surface via hydrophobic interactions. Then, moving the liquid-air interface along the substrate will drag the hydrated part of DNA molecules and result in stretching of the molecules with one or both ends attached to the surface. While this is the basic idea, different combing techniques use different strategies to induce the moving interface. The force applied by the receding meniscus is typically estimated ~400 pN, which is two orders of magnitude greater than the entropic forces keeping DNA in its coil structure (and comparable to the bond strength between biotin and streptavidin), yet not enough to detach DNA from the surface [15, 16]. This procedure results in formation of uniformly stretched and parallel arrays of DNA molecules on the substrate. Movement of the meniscus can be induced by many different techniques such as by pulling a substrate out of solution containing DNA or evaporating a droplet as will be discussed below.

Before discussing various combing procedures, an important point should be made. When the group lead by David (physicist) and Aaron (molecular neurobiologist) Bensimon, who contributed significantly to different aspects of combing, introduced molecular combing in 1994 [16], the technique relied on hydrophobic interactions of partially melted DNA ends with the surface. Different hydrophobic surfaces have been employed for combing: glass or silicon surfaces modified with hydrophobic silanes [17], polystyrene surfaces [18], polymethylmetacrylat (PMMA) surfaces [5], polydimethylsiloxane (PDMS) [19], etc. However, the term "combing" has also been applied to the experiments performed on charged surfaces ranging from clean glass [20] to the surfaces modified with positive coatings including amino terminated groups such as polylysin and polyhystidine [20], 3-aminopropyltrimethoxysilane (APTMS)-coated [21] and 3-aminopropyltriethoxysilane (APTES)-coated surfaces [20], where electrostatic interaction of the DNA molecule with the surface plays important role. It should be noted that typically the use of charged substrates results in unspecific electrostatic interaction of DNA mid-segments with the surface, while the hydrophobic surfaces at proper values of pH and ionic strength ensure more specific adsorption of DNA from extremities via relatively strong hydrophobic interactions.

Later on, with attempts to achieve further control over positioning of DNA molecules, the methods were introduced that involved a combination of combing with lithographic approaches [18, 22]. This has resulted in development of methods with more capabilities and, therefore, a wider range of potential applications in nanotechnology and biomedicine. In this context, in addition to aligning DNA molecules on the substrate, controllable and reproducible patterning of arrays of aligned DNA molecules with a desired distance becomes possible [22].

The primary objective of this review is to focus on the main experimental aspects of molecular combing of DNA, including the choice of surface, pH and ionic strength of the solution as well as to discuss various techniques useful for positioning and aligning DNA molecules. While the applications of combing in genomics, DNA replication and cancer studies have been recently reviewed by John Herrik and Aaron Bensimon [23], no comprehensive review has been published on molecular combing of DNA as a technique with a variety of applications from biomedicine to nanoelectronics. Here we will attempt to cover most of significant achievements in nanoscience that involved application of combing technique. To meet this end, the underlying mechanism of combing will be described first, followed by the discussion of different techniques in molecular combing. The third part of the review will focus on main applications of combing in nanotechnology and biomedicine.

2 Mechanism

Combing occurs in three main steps: adsorption of the ends of floating coiled DNA molecules to a substrate, stretching DNA by the forces exerted by the receding meniscus, followed by relaxation of the deposited DNA on the substrate to its final length. A simple procedure for combing DNA is presented in Figure 1.

As discussed before, under certain pH and ionic strength conditions, DNA is partially melted at the ends. Upon unwinding of DNA at the ends, the hydrophobic core of DNA double helix is exposed and is readily attracted to the hydrophobic surfaces. When the moving air-water interface is passing along the solid substrate, the coiled DNA is stretched with one or both ends attached to the surface. At very low pH values (typically, below pH 4), DNA tends to denature along the length of the molecule resulting in non-specific binding to the surface at several random points along the DNA length. This leads to

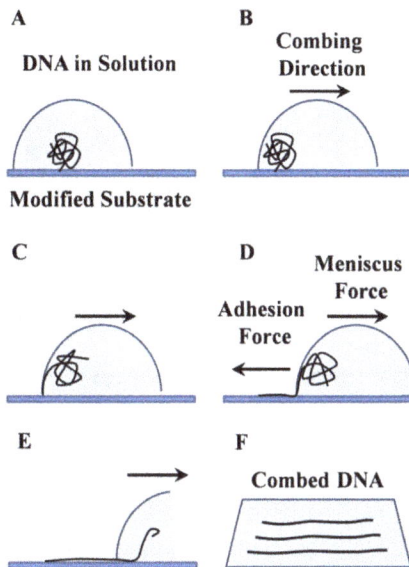

Figure 1 Schematic representation of basic steps in molecular combing of DNA. Under certain conditions of pH and ionic strength, DNA is partially melted at the extremities and then adsorbed to the surface by the ends. When the moving meniscus is formed, the adhesion force between DNA and the surface dominates the meniscus force, resulting in random coiled DNA being stretched, while the meniscus passes along the surface.

low and non-uniform DNA stretching. On the other hand, at physiological pH values, the unwinding at the extremities is inhibited and most molecules leave the surface. Although the optimal pH range depends on the nature of the surface, structure of DNA extremities (e.g., presence of sticky ends) and ionic strength, typically, the optimal unwinding of DNA extremities is achieved in a narrow range of pH between 5 to 5.6 [24].

Combing process is governed by the surface tension forces existing at the air-water interface. During movement of the meniscus, the receding air-water interface leaves the bound DNA molecules fully extended behind and stretched on the dry substrate. This means that the surface tension of the air-water interface at each point exhibits a force on the DNA on that point. This force, F, is proportional to the surface tension γ at the air-water interface and the wetted diameter d of the DNA molecule: $F = \gamma \pi d$. In this equation we assume that first, the DNA chain is completely wet by the water and second, the DNA chain is stretched perpendicular to the air-water interface. Taking

$\gamma = 7 \times 10^{-2}$ N/m for air/water interface, and $d = 2.2$ nm for B-DNA, the force acting on DNA can be estimated as 400 pN. This force is two orders of magnitude greater than the entropic forces that maintain DNA in its native random coil structure, yet smaller than the force required to break the covalent bonds in DNA backbone, which is of the order of 1 nN [16]. As was established by Smith et al. [25], when the force acting on DNA increases from the entropic range to a level from 6 to 60 pN, DNA starts to behave as an elastic rod with a stretch modulus of $E \cdot A \sim 1000$ pN where E is the Young modulus of B-DNA and A its effective cross-sectional area [26]. In this force range, the force vs. extension curve follows Hook's law:

$$F = E \cdot A(l/l_0 - 1),$$

where $E = 1.1 \times 10^8$ N/m^2 is the Young modulus of DNA molecules, $A = 3.8 \times 10^{-18}$ m^2 is the cross-sectional area of a molecule, l_0 is the contour length of DNA, and l/l_0 is the relative extension. When the force reaches approximately 70 pN, DNA suddenly yields and overstretches up to 1.7-fold its B-form crystallographic length [25]. This corresponds to a transition from the B-form DNA to an "overstretched" DNA that believed to be a "ladder" composed of two parallel ssDNA strands. At even higher forces the dsDNA melts into two parallel ssDNA strands, which can be further stressed without breaking up to 800 pN [27, 28]. Unless there are nicks in the DNA strands, the overstretching transition is reversible and the DNA will return to its B-form upon releasing the stress. On the other hand, the melting transition is loading rate dependent and is not expected to occur during fast DNA stretching characteristic for combing process [19].

As illustrated in Figure 1, when the moving meniscus is formed, there is a constant stretching force parallel to the direction of combing, which acts on DNA molecules in the vicinity of the contact line. If the adhesion force between DNA and the surface is sufficiently high and dominates the meniscus force, the DNA undergoes transition from coil to the stretched form [19, 29]. This condition puts a lower limit on the value of the adhesion force holding the extremities of DNA.

The last stage of combing process corresponds to the DNA molecule relaxing on solid surface after passage of the meniscus. The DNA undergoes relaxation towards its B-form until the stress in DNA is balanced by its interaction with the surface (van der Waals, hydrophobic, etc.). Typically this corresponds to the relative extension of 1.2 times, although higher values are sometimes reported [21].

It has been shown that the density of combed molecules on the surface is not heavily affected by increasing the DNA concentration in solution; rather, it is strongly dependent on the radius of gyration R_G[1] of the DNA molecule in solution, which essentially limits how closely DNA molecules can approach each other [18]. Although this size can be affected by ionic strength of the solution [19], in order to significantly increase the density of combed DNA, it might be necessary to repeat combing procedure for a number of times [18].

Studies suggest that combing is an irreversible process, which means that if the substrates with combed DNA has been dried sufficiently, the already-combed DNA molecules remain stretched on and attached to the surface upon rehydration and it is possible to perform combing repeatedly [15, 18]. The physics and theoretical aspect of DNA elastisicity and stretching is beyond the aim of this review and the readers are encouraged to follow the extensive discussions by Bensimon and his team [15, 26, 30].

3 Basic Combing Techniques

3.1 The Original Bensimon Method

In the original Bensimon method, a droplet containing DNA solution (pH = 5.5) was deposited on a silanized coverslip covered with another glass and left to evaporate. The movement of the air/water interface occurred upon evaporation of the solution lead to DNA stretched on the silanized glass perpendicular to the meniscus (Figure 2) [15, 20]. In 2004, Zheng et al. used the same method for combing of DNA on cetyltrimethyl ammonium bromide (CTAB)-coated surface instead of silane-coated surface. CTAB is a cationic surfactant which is also used as a germicidal chemical due to its ability to bind to DNA. The stretched DNA molecules were reported to be 30% longer compared to the contour length of λ-DNA at pH = 5.5 [31].

3.2 Moving Coverslip

A variation of Bensimon's method was used three years later by Yokota et al. based on mechanical moving of the meniscus by a droplet-spreading apparatus (Figure 2). After incubation of DNA solution which is spread on a

---

[1] R_G – the half average size of the coiled DNA molecule in solution – is a widely used measure for the characterization of the configuration of DNA as a polymer. It measures the root-mean square distance of the collection of segments from their common center of mass.

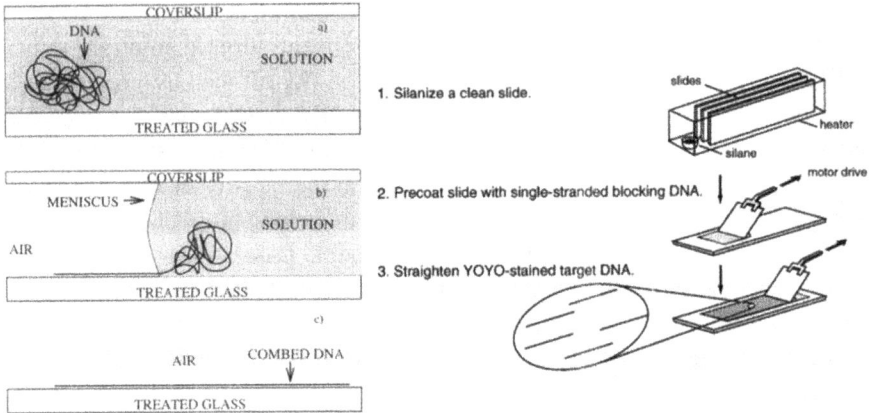

Figure 2 Left: The original method – proposed by Bensimon – for molecular combing of DNA. During incubation, DNA molecules become attached to silanized glass by their extremities. When the upper coverslip is removed, molecules are uniformly stretched and aligned by the receding air-water meniscus (left). Reprinted with permission from [20]. © 1997 Elsevier. Right: Variation of Bensimon's method based on moving of the meniscus by a motor-driven apparatus. Reprinted with permission from [32]. © 1997 Oxford University Press.

silanized glass slide, the coverslip is dragged across the slide surface using a motor-driven apparatus at a constant speed [32].

3.3 Dynamic Combing

In 1997, Michalet et al. reported dynamic molecular combing for stretching total genomic DNA based on Langmuir-Blodgett deposition technique [33]. In this method, after preparation of trichlorosilane-coated surfaces, the substrates were incubated in the solution of either yeast or human genomic DNA (pH = 5.5) for 5 min, after which the surface was pulled out of the solution at a constant speed, as shown in Figure 3. When the cover slip was pulled out of the solution, the anchoring points moved upward together with the surface, while the fixed horizontal meniscus exerted a constant downward vertical force. An important advantage of dynamic combing procedure over the original Bensimon method is that the use of large volumes (2–20 ml) of buffer solution in this method allows better control of pH of DNA solution compared to small volumes of 5 μL. This is particularly important considering the narrow pH range in which molecular combing is observed.

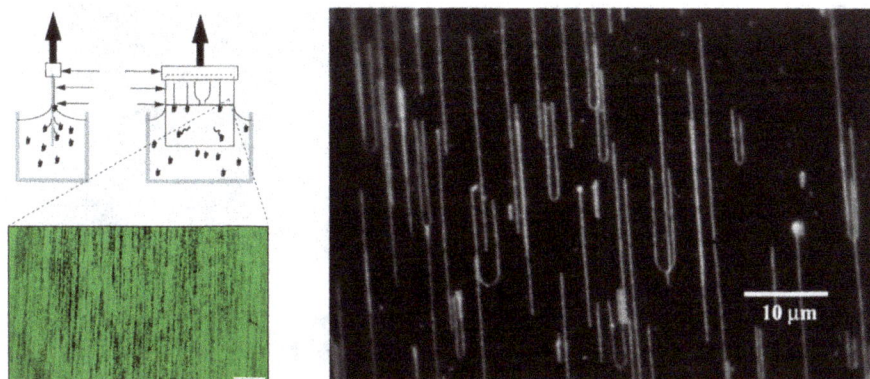

Figure 3 Dynamic combing employed for stretching large quantities of genomic DNA. Left: A silanized coverslip is incubated in DNA solution in order to allow adsorption of DNA molecules to the surface. The coverslip is then pulled out of the solution in vertical direction leaving up to several hundred genomic DNA molecules behind. Reprinted with permission from [33]. © 1997 American Association for the Advancement of Science. Right: Fluorescence microscopy image of the DNA strands aligned using the dynamic molecular combing method. Reprinted with permission from [34]. © 2003 Elsevier.

3.4 Combing with a Moving Droplet

Probably the simplest method for combing DNA on the surface is stretching by gas flow. In this method, a droplet containing DNA solution is deposited and incubated on modified substrate and the sample is blown dried with nitrogen gas. If the gas is blown in appropriate direction and angle ($\sim$45° relative to the surface), DNA molecules will be stretched. The method was first proposed in 1998 [35] and new variations of the method were used later [36]. The disadvantage of this method is that the actual parameters of the gas flow strongly depend on the distance between the surface and the gas source and may vary between the experiments.

Combing DNA using gravitational force is a common method too [37, 38]. In this approach, the substrate is tilted approximately 80° with respect to its horizontal position and a droplet of DNA solution is placed at the upper edge of the substrate. The droplet slides down as a result of gravitational force and the moving air-water interface stretches DNA molecules [19].

3.5 Combing by Droplet Suction

In 2002, Nakao et al. proposed a procedure for generating highly-aligned long strands of DNA [39]. The cover slips were coated with poly(vinylcarbazole)

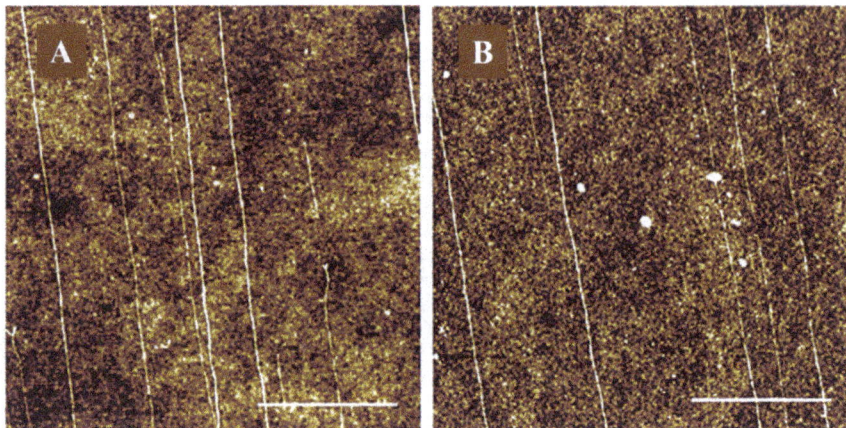

Figure 4 AFM topography of DNA stretched on PPhenaz-coated glass (A) and PVCz-coated glass using pipette sucking method. The white bar denotes 1.0 μm (B). Reprinted with permission from [39]. © 2002 American Chemical Society.

(PVCz) and polyphenazasiline (PPhenaz) polymers via spin-coating. A droplet of DNA solution was then deposited and incubated on polymer-coated surfaces. Combing of DNA was performed by sucking the droplet up by a pipette which induced the movement of the air/water interface and subsequent alignment of DNA along the central direction of the droplet. It was also possible to comb DNA on other substrates such as silicon wafers and graphite (Figure 4) [39].

Later on in 2006, the same procedure was used for combing DNA on a surface of a di-block copolymer; polystyrene-b-poly(methyl methacrylate) (PS-b-PMMA) [40].

4 Complex Combing Methods

While initially the combing techniques were designed to simply achieve DNA stretching on non-patterned, flat surfaces mostly for applications in genomic analysis, application of DNA in molecular electronics requires better control over the orientation; position and distance between combed molecules. These methods generally involve a combination of combing with micro- and nano-fabrication techniques such as lithographic patterning, Dip Pen Lithography (DPN) and Transfer Printing (PT).

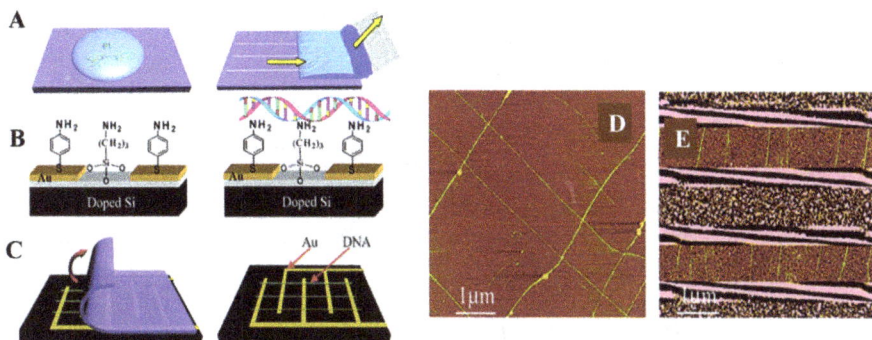

Figure 5 Combing DNA on Si substrate (A) Molecular combing was used in order to achieve aligned DNA strands using a piece of filter paper. (B) Si substrate was modified by APTES producing an amino-terminated surface to anchor DNA molecules. (C) Combed DNA molecules are transferred from PDMS to the Si chip by contact printing. (D) DNA crossed structures were prepared with layer-by-layer contact printing. (E) DNA combed across electrodes on Si substrates. Reprinted with permission from [41]. © 2005 American Chemical Society.

4.1 PDMS Printing

In 2005, Zhang et al reported a combination of molecular combing and transfer printing of λ-DNA, in which a piece of filter paper was used to control the movement of the meniscus [41]. This technique involved incubation of a drop of λ-DNA solution (pH = 8) on PDMS substrate followed by adsorption of the solution at a constant rate with a piece of filter paper, as shown in Figure 5. Afterwards, the combed DNA strands were transferred from the PDMS sheets onto APTES-treated Si surface by contact printing. In this way, using layer-by-layer contact printing, it was possible to form aligned patterns of DNA strands on Si surface, e.g., crossing each other or aligned across micro electrodes. The PDMS printing method has the advantage of uniformity of the stretched molecules, simplicity, reproducibility and the fact that the speed of the meniscus is controlled by the pore size of the filter paper [41].

4.2 Combing across Nanoelectrodes

In a recent approach, the authors have developed a method for combing DNA on Si substrates bearing lithographically-patterned Pt/Cr nanoelectrodes [17]. In this method, Si substrates were silanized using gas phase deposition of N-Octyldimethylchlorosilane. Compared to the commonly used N-Octadecyltrichlorosilane, the use of N-Octyldimethylchlorosilane resulted in a molecularly-smooth surface with an average contact angle of ∼90° which made it easy to drag the droplet on the modified surface. After deposition and incubation of DNA on silane-modified substrate, the capillary forces between the droplet and the plastic pipette tip was used to gently drag the droplet out of the surface. The movement of the air/water interface results in highly-ordered alignment of DNA molecules that were stretched up to 160% of their original crystallographic length. The average percentage of stretching was calculated as 122%. Additionally, using the relatively large meniscus force produced by this method, it was also possible to comb more rigid derivatives of DNA such as DNA-peptide nanowires [17, 42]. These structures were composed of a DNA core, coated with a peripheral layer of self-assembled short cationic peptides [43–46]. While combing of these structures was not efficient using other available techniques, it was possible to comb these nanowires to their full length, as shown in Figure 6C.

The technique was also used for combing across silicon substrates with nano-fabricated platinum electrodes. In that case, it was possible to achieve 700 nm-long stretches of dsDNA across nanoelectrodes [17].

4.3 Combing on E-Beam Patterned Polystyrene Templates

In 2001, Klein et al. developed a new technique based on stretching of DNA between micro-fabricated polystyrene lines. Polystyrene readily forms cross links between chains under e-beam irradiation and can be used as negative e-beam resist. Since cross-linked chains are less soluble in organic solvents compared to the initial material, and DNA binds equally well to cross-linked polystyrene, it was possible to control the binding regions on the substrate and achieve positioning of combed DNA molecules [18]. The Si/SiO_2 substrates with e-beam patterned polystyrene layer were then dipped into a solution of λ-DNA for approximately 1 min to allow hydrophobic interactions between DNA extremities and polystyrene lines. Next, the silicon substrate was pulled out with a constant speed which resulted in stretching of DNA across and between the polystyrene lines (Figure 7) [18].

The advantage of this method is that it is possible to achieve control over positioning of DNA arrays on the substrate. This offers a potential advantage in studies on DNA nanowires and DNA-based nanoscale devices. Additionally, the technique offers a good possibility of large-scale automated fabrication of DNA nanowires for applications in either DNA electronic or population-based genetic diseases and genomic screening.

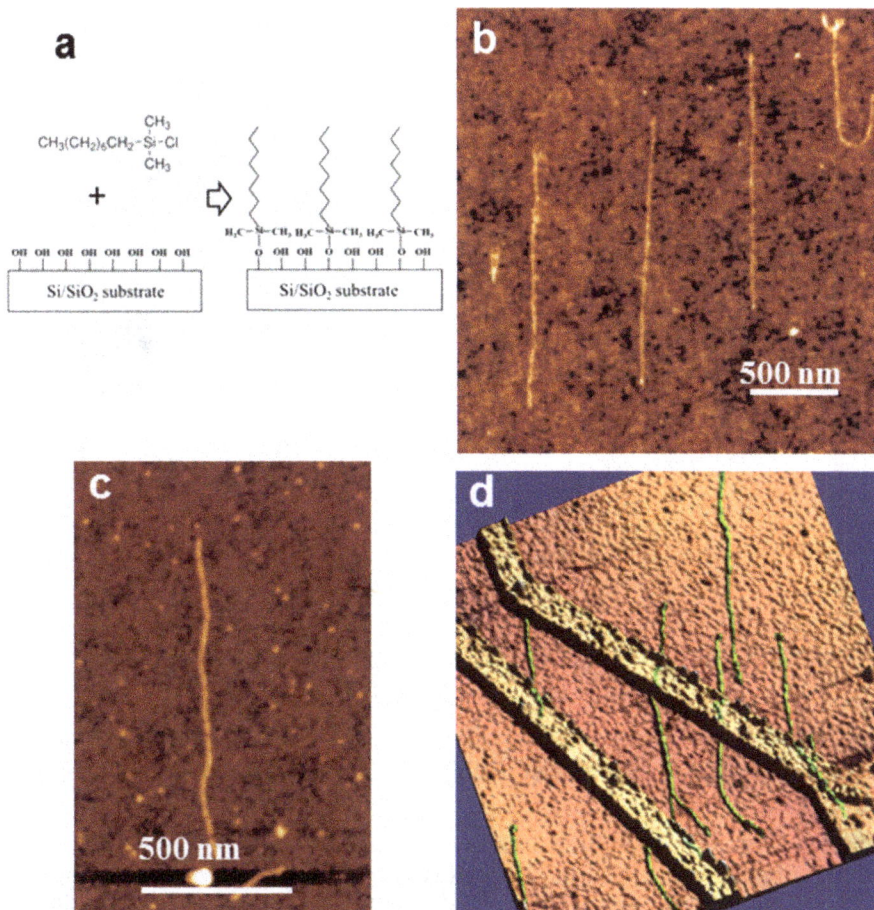

Figure 6 Combing DNA molecules using pipette tip method (a). Schematics of surface modification with Octyldimethylchlorosilane. The molecules can form only a single bond to the surface, therefore producing smooth monolayers. (b) combed dsDNA molecules on the surface, average height ~0.7 nm (c) combed peptide-coated DNA, average height 3.4 nm (d) dsDNA molecules combed across nano-fabricated electrodes ($2 \times 2 \mu$m scan area) [17].

4.4 Combing on Dip Pen Lithography (DPN) Templates

A combination of molecular combing and Dip Pen Lithography (DPN) is another example of how one could achieve more control over DNA molecule by using a combination of combing with other procedures. In the method described by Nyamjav and Ivanisevic in 2003, different patterns with the size down to 50 nm were "inked" using contact-mode AFM scanning of

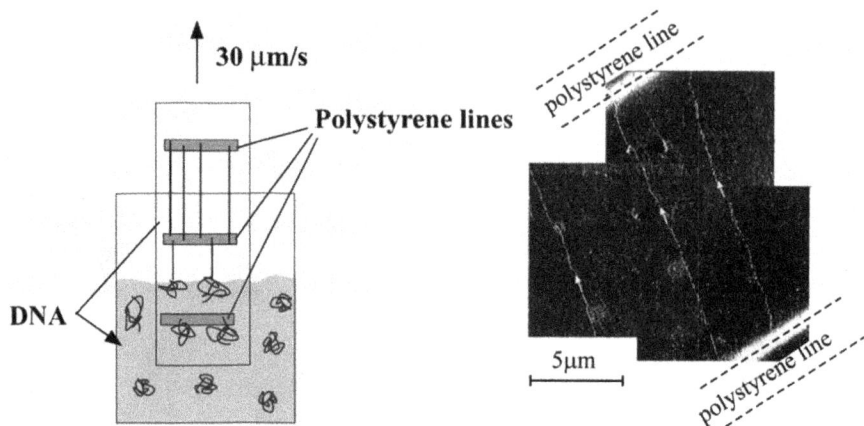

Figure 7 Combing across polystyrene lines microfabricated on Si/SiO$_2$ substrates. The substrate is dipped into DNA solution and then retracted at a constant speed. DNA molecules bind to the polystyrene lines and are stretched by the meniscus force (left). Amplitude AFM image of DNA stretched between patterned polystyrene lines (right). Reprinted with permission from [18]. © 2001, American Institute of Physics.

poly(allylamine hydrochloride) (PAH)-coated tip on Si substrate [47]. Molecular combing of DNA was then performed on PAH-patterned templates and DNA strands were combed along the PAH layer. Using this method, surface templates with well-defined regions of positive and negative charges has been demonstrated that could be used for guided deposition and combing of DNA together with other biomolecules [47]. Two years later, the same team extended this technique to achieve templating of DNA with magnetic nanoparticles. In this technique the DNA molecules were templated by Fe$_3$O$_4$ nanoparticles prior to combing on PAH layer. In this way, it was possible to achieve magnetic nanowires positioned and aligned on the substrate [48].

4.5 Transfer Printing of Combed DNA

In 2003, Gad et al. used the combing procedure described in Section 3.5, for stretching λ-DNA on PDMS sheets, followed by transfer-printing of DNA from PDMS to mica substrate [49]. DNA molecules were initially combed on PDMS by both methods of pipette-sucking [49] and pulling up the substrate out of DNA solution [50], and then were micro-contact printed on freshly cleaved mica resulting in stretched DNA patterns on mica surface. The

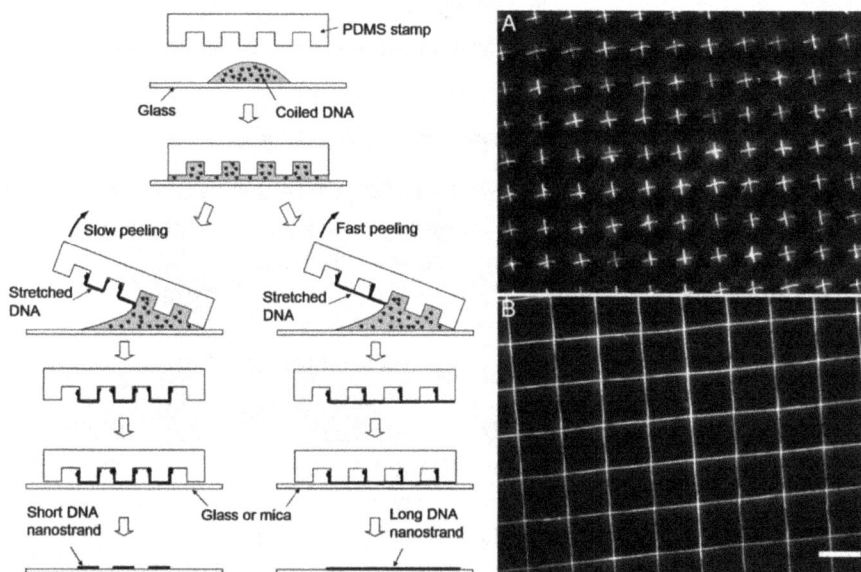

Figure 8 Procedure for transfer-printing of DNA. After incubation of DNA on a glass cover-slip, a PDMS stamp is pressed on the solution, and peeled off the surface from one end with the other end remaining in contact with the glass surface (left). It results in generation of ordered arrays of either short (A, slow peeling) or long (B, fast peeling) patterns of stretched DNA on mica or glass substrate. Reprinted with permission from [22]. © 2005 National Academy of Sciences, USA.

method allows creation of complex patterns by repeated printing; however, the exact position and length of DNA molecules could not be controlled.

Two years later, Guan and Lee combined molecular combing with litho-graphic approaches in order to generate ordered alignments of DNA on substrate [22]. In their method, the droplet of DNA solution was first incub-ated on a glass coverslip, and a PDMS stamp was then placed on top of the solution. After manual peeling from one end, the stamp was transferred on a new solid surface by direct contact printing which literally resulted in short (peeling with low-speed) or long (peeling with high-speed) arrays of DNA on the surface. The advantage of this technique is that more complex patterns could be produced by additional steps of printing (Figure 8) [22].

Later on, more techniques were developed based on using capillary as-sembly in combination with soft-lithography in order to achieve ordered arrays of isolated DNA strands. Some of these techniques resulted in stretch-ing of DNA up to 160% of its contour length [21]. We would mention in particular the procedures involving UV-light for end-specific binding of DNA to the patterned thin, spin-coated aminoterpolymer (ATP) film [51].

5 Parameters Affecting Combing Efficacy

As discussed above, the extent of stretching depends on a number of combing parameters such as the choice of substrate, buffer, surface functionalization method, etc. In addition, in nanoelectronic applications, the control over the position and orientation of DNA strands on the substrate is of great importance.

It has been shown that change in pH has a significant effect on both density of DNA molecules adsorbed on the surface and the extent of stretching. The optimal pH for combing DNA on a surface is highly dependent on the nature of the surface used. In general, most results, if not all of them [52], agree that for hydrophobic surfaces, pH values in the range from 5–5.5 are ideal for adsorption of DNA to the substrates [17, 20, 31]. At very low pH values DNA bases are largely protonated (e.g., pH = 3 and 0.1 M salt content corresponds to 50% protonation) [20]. This protonation weakens hydrogen bonds holding the strands together, decreases the melting temperature, and partially exposes the hydrophobic core of DNA helix hence causing nonspecific adsorption of DNA to the surface. However, with the increase in pH value, the melting occurs less frequently, so at pH 5.5, only the extremities of DNA are sufficiently hydrophobic to bind to a hydrophobic surface [20].

It has also been reported that the efficacy of combing is dependent on the ionic strength of the buffer used. For surfaces such as PDMS, increasing the concentration of NaCl in the buffer up to 100 mM, resulted in better stretching of DNA where the number of stretched DNA molecules reached a maximum [19]. In general, the ionic strength of the solution controls screening of the DNA backbone and substrate charges and as such can affect a number of parameters important for combing, including melting temperature and melting extent at the extremities, gyration radius of DNA coils, as well as DNA interaction with the surface.

The choice of surface functionalization method is also important. For instance, in silanization procedure, the nature of the silane used defines the extent of hydrophobicity of the substrate which directly affects the strength of interaction of DNA with the substrate. While the use of silanes have been a common choice for preparation of hydrophobic surfaces, some studies showed that a number of polymers containing π-conjugation units (such as polyphenazasiline, PPhenaz and poly(vinylcarbazole), PVCz) also result in highly-aligned DNA molecules [39]. In general, the degree of stretching is usually higher in hydrophobic rather than hydrophilic surfaces, due to strong and specific adsorption of DNA and higher meniscus forces (>100 pN) [20, 31]. Effect of hydrophobicity of the substrate on combing quality is discussed in [52].

The speed of the moving meniscus also affects the quality of combing. This could be controlled by either using different sizes of a filter paper or a motor-driven apparatus [18], as discussed above. Generally, meniscus speed in the range of 300–500 μm/s is acceptable [33, 41], but studies suggest that it could be also as high as 5 mm/s when using a filter paper with a pore size of 11 μm to absorb the DNA solution [41]. It has also been shown that increase in the adsorption speed and hence the receding meniscus speed, results in decrease in the density of DNA. However, significant reduction in meniscus speed (down to 250 μm/s) also decreased the density of stretched DNA [17, 41]. In addition, both the direction of blow drying and the angle at which the nitrogen gas is blown to the surface are important. The best results are achieved when the direction of gas flow has a tilt angle of 45° from that of the surfaces. Blowing at a large angle (parallel with the surface) causes DNA molecules to fly out of the surface, while blowing at too small an angle (perpendicular to the surface) would not stretch DNA molecules in a single direction [35]. The extent of stretching is affected by the length of DNA but is sequence-independent. In addition, the concentration of DNA as well as the incubation time of DNA needs to be optimized.

It should also be noted that the density of molecules stretched on the substrates does not depend on the DNA concentration; rather, it is related to the radius of gyration RG of the DNA molecule in solution. Therefore, the best way to increase the density of combed DNA molecules is probably to repeat the combing procedure few times until the desired density is achieved [18].

A comprehensive study of the parameters that affect the efficiency of DNA combing could be found in [29].

6 Applications

Most interestingly, molecular combing of DNA is used as a tool in many fields of science as diverse as physics, biology, and medicine. These applications fall into one of these categories:

6.1 A Tool in Genomic Studies

In fact, the early advances in DNA molecular combing were exclusive to high-resolution studies of chromosomal DNA molecules and studying genetic rearrangements. The first genetic disease diagnosed with the help of combing was previously-undetectable micro-deletions in the TSC2 gene responsible for the disease tuberous sclerosis [23, 33]. Soon after, combing was adapted for many more applications such as detection of other genetic disorders and cancer. For instance, alterations in BRCA1 and BRCA2 are responsible for the majority of breast cancers; however, most previous PCR-based methods for detection of these alterations were not capable of detection of large re-arrangements in genome. The use of molecular combing of DNA in such studies was significantly helpful in overcoming this problem. For instance, Gad et al. studied large rearrangements in BRCA1 using a combination of combing with a system of four-color bar coding. The color bar code method was able to detect micro-deletions on the order of 10 kb in a variety of BRCA cell lines from patients with breast cancer [53].

Fluorescence *in situ* Hybridization (FISH) is a cytogenetic technique used for detection of specific DNA sequences on chromosomes. In this technique, fluorescently labeled probes are used to hybridize to the desired sequences on chromosome. Fluorescence microscopy can then be used to visualize the fluorescent signals emitted by the fluorescent probe on the chromosome [5, 54]. Dynamic molecular combing has been a common choice for combing of DNA, with the possibility to comb total genomic DNA into a high density array of well-separated molecules (typically 50–100 diploid human genomes with lengths between 200–600 kb) [33]. After combing, regions of interest are visualized and characterized by fluorescence microscopy. Combing of DNA is also used in fiber-FISH in order to stretch large amounts of DNA (bacterial and yeast expression vectors or even whole genomes) on silanized glass for high-resolution detection of genes [54]. It is also possible to determine the ordering, orientation and distance between genes, as well as the existence of certain genomic rearrangements (e.g. deletions) [55].

The main advantage of molecular combing in genomic studies is the possibility to comb high concentrations of total genomic DNA in conditions preserving it from excessive shearing. For instance, it is possible to prepare high density of combed fibers of total human genome up to hundreds of long fragments (longer than several hundred kilo-bases) per cover slip [33]. Moreover, it is possible to prepare a very large number of identical substrates at the same time [5].

Application of combing technique in genetics gradually evolved to a method for studying dynamics of the replication of the genome [38, 56–59]. A comprehensive review on application of molecular combing in genomic studies can be found in [23].

6.2 Nanoelectronics

Application of DNA as a building block in future nanoelectronic devices is a charming idea. One of the main challenges in design and development of nanocircuits is inter-element wiring and positioning of nanodevices at nanoscale [60]. In this context, DNA is an attractive candidate due to self-assembly and recognition features which are unique to this molecule [7–9]. Taking advantage of these properties, it is possible to design procedures for controlled wiring of nanocircuits using DNA nanowires with adhesive ends. In addition, fabrication of multi terminal junction devices by crossing nanowires would be possible using branched DNA junctions (Holiday junctions) [61, 62]. Based on these reasons, and even though the electrical conductivity of DNA has been a subject of strong debates in last decade [63–71], DNA is still a unique candidate as a potential nanowire and a subject of many ongoing investigations [72]. In addition to normal dsDNA, modified forms of DNA including M-DNA (a complex of B-DNA with divalent metal ions Co^{2+}, Ni^{2+} and Zn^{2+}) [73], metal-coated DNA [74, 75], guanine quadruplexes G4-DNA [84], and DNA-peptide wires [42] have also been discussed as future nanowires.

Molecular combing of DNA plays an important role in such experiments since all these conductivity measurements primarily require proper stretching of DNA across electrodes. Therefore, combing method has been used as a common choice of a convenient yet effective method with no need for any prior modification to DNA [37, 63, 76, 77]. In some cases, combing of DNA has been used in combination with other lithographic techniques [51] or in conductivity measurement of overstretched DNA molecules [78]. Molecular combing has also been used to comb DNA derivatives such as DNA-peptide conjugates [44–46, 79] across nanoelectrodes for either analyzing DNA-protein interaction [80] or conductivity measurements [42].

As mentioned earlier, DNA-templated nanowires have been a subject of many studies in the last decade. The reason is that the molecular recognition and self-assembly properties of DNA make it a suitable potential template to organize and interconnect assemblies of nanoparticles on surfaces [81]. DNA-templated nanowires consist of DNA as the template and either metallic or semi-conductive nanoparticles as the coating layer. Such nanomaterials are excellent candidates for future nanoelectronics due to the fact that they possess structural advantages of DNA as a nanowire (self-assembly and recognition) together with the desired conductivity of metallic or semi-conductive particles.

The preparation of arrays of DNA-templated nanowires also include the use of combing technique for immobilization and stretching of DNA prior to assembly of nanoparticles on DNA template. For instance, Nakao et al. developed a method for ordered assembly of gold nanoparticles (Au-NPs) along DNA molecules [81]. Molecular combing has been also used in combination with dip pen lithography in order to stretch DNA molecules pre-templated with Fe_3O_4 magnetic nanoparticles [48].

In a more interesting approach, Keren et al. used combing of DNA to achieve DNA-templated nanowires in a system of nano-bio-lithography where desired sites on DNA are masked via homologous recombination by RecA protein, while other parts are coated with metallic nanoparticles in order to generate molecularly accurate DNA-templated junctions [82]. In a more recent approach, the same team used molecular combing technique as part of the fabrication procedure for DNA-templated nanofield effect transistors [83].

7 Conclusions

The emergence of new methods for nano-scale investigation and manipulation is very promising. Among these methods, molecular combing of DNA has shown promise in bridging between organic and inorganic elements. The large body of literature and the many experiments that involve combing is the best indication of the reliability and efficiency of this method. Besides biomedical applications in cytogenetics and cancer genomics such as screening of genome for cancer and genetic diseases due to genomic rearrangements, the application of molecular combing has allowed many state-of-the-art advances in new areas of research in nanophysics and nanoelectronics. Development of DNA-templated nanowires for future nanoelectronics, single molecule lithography and nanofabrication of DNA-based nanodevices are good examples that were discussed in this review. The simplicity of the combing technique and its broad range of application on one hand and rich physics of DNA deformation involved in the combing process on the other hand warrant that this review will not stay complete or comprehensive for long, but hopefully will rather encourage the use of this elegant technique.

Acknowledgements

The authors gratefully acknowledge support by EU COST action MP0802 "Self-assembled guanosine structures for molecular electronic devices" and grants from the Obel Family Foundation.

References

[1] L. B. Kish, Phys. Lett. A, 305, 144–149 (2002).
[2] M. Lundstrom, Science, 299, 210–211 (2003).
[3] A. M. van Oijen, J. J. Loparo, Annu. Rev. Biophys., 39, 429–448 (2010).
[4] A. Stewart, Mol. Med. Today, 4, 2 (1998).
[5] J. Herrick, A. Bensimon, Chromosome Res., 7, 409–423 (1999).
[6] F. Zamora, M. P. Amo-Ochoa, P. J. Sanz Miguel, O. Castillo, Inorg. Chim. Acta., 362, 691–706 (2009).
[7] C. M. Niemeyer, Curr. Opin. Chem. Biol., 4, 609–618 (2004).
[8] R. Chhabra, J. Sharma, Y. Liu, S. Rinker, H. Yan, Adv. Drug Deliver. Rev., 62, 617–625 (2010).
[9] E. Farjami, L. Clima, K. Gothelf, E. E. Ferapontova, Anal. Chem., 83, 1594–1602 (2011).
[10] C. G. Baumann, V. A. Bloomfield, S. B. Smith, C. Bustamante, M. D. Wang, S. M. Block, Biophys. J., 78, 1965–1978 (2000).
[11] H. Yokota, J. Sunwoo, M. Sarikaya, G. van den Engh, R. Aebersold, Anal. Chem., 71, 4418–4422 (1999).
[12] L. M. Bellan, J. D. Cross, E. A. Strychalski, J. Moran-Mirabal, H. G. Craighead, Nano Lett., 6, 2526–2530 (2006).
[13] L. J. Guo, X. Cheng, C.-F. Chou, Nano Lett., 4, 69–73 (2004).
[14] D. J. Doolittle, G. Muller, H. E. Scribner, Chem. Toxic., 25, 399–405 (1987).
[15] D. Bensimon, A. J. Simon, V. Croquette, A. Bensimon, Phys. Rev. Lett., 74, 4754–4757 (1995).
[16] A. Bensimon, A. Simon, A. Chiffaudel, V. Croquette, F. Heslot, D. Bensimon, Science, 265, 2096–2097 (1994).
[17] Z. Esmail Nazari, L. Gurevich, Beilstein J. Nanotechnol., in press (2013).
[18] D. C. G. Klein, L. Gurevich, J. W. Janssen, L. P. Kouwenhoven, J. D. Carbeck, L. L. Sohn, Appl. Phys. Lett., 78, 2396–2398 (2001).
[19] A. Benke, M. Mertig, W. Pompe, Nanotechnology, 22, 035304 (2011).
[20] J. F. Allemand, D. Bensimon, L. Jullien, A. Bensimon, V. Croquette, Biophys. J., 73, 2064–2070 (1997).
[21] A. Cerf, C. Thibault, M. Geneviève, C. Vieu, Microelectron. Eng., 86, 1419–1423 (2009).
[22] J. Guan, L. J. Lee, PNAS, 102, 18321–18325 (2005).
[23] J. Herrick, A. Bensimon, Methods Mol. Biol., 521, 71–101 (2009).
[24] J. Herrick, A. Bensimon, Biochimie, 81, 859–871 (1999).

[25] S. B. Smith, Y. Cui, C. Bustamante, Science, 271, 795–799 (1996).

[26] T. Strick, J. Allemand, V. Croquette, D. Bensimon, Prog. Biophys. Mol. Biol., 74, 115–140 (2000).

[27] M. Rief, H. Clausen-Schaumann, H. E. Gaub, Nat. Struct. Biol., 6, 346–349 (1999).

[28] H. Clausen-Schaumann, M. Rief, C. Tolksdorf, H. E. Gaub, Biophys. J., 78, 1997–2007 (2000).

[29] J. H. Kim, W.-X. Shi, R. G. Larson, Langmuir, 23, 755–764 (2007).

[30] T. R. Strick, J.-F. Allemand, D. Bensimon, V. Croquette, Biophys. J., 74, 2016–2028 (1998).

[31] H.-Z. Zheng, D.-W. Pang, Z.-X. Lu, Z.-L. Zhang, Z.-X. Xie, Biophys. Chem., 112, 27–33 (2004).

[32] H. Yokota, F. Johnson, H. Lu, R. M. Robinson, A. M. Belu, M. D. Garrison, B. D. Ratner, B. J. Trask, D. L. Miller, Nucleic Acids Res., 25, 1064–1070 (1997).

[33] X. Michalet, R. Ekong, F. Fougerousse, S. Rousseaux, C. Schurra, N. Hornigold, M. van Slegtenhorst, J. Wolfe, S. Povey, J. S. Beckmann, A. Bensimon, Science, 277, 1518–1523 (1997).

[34] K. J. Kwak, S. Yoda, M. Fujihira, Appl. Surf. Sci., 210, 73–78 (2003).

[35] J. Li, C. Bai, C. Wang, C. Zhu, Z. Lin, Q. Li, E. Cao, Nucleic Acids Res., 26, 4785–4786 (1998).

[36] Z. Deng, C. Mao, Nano Lett., 3, 1545–1548 (2003).

[37] T. Heim, T. Mélin, D. Deresmes, D. Vuillaume, Appl. Phys. Lett., 85, 2637–2639 (2004).

[38] M. Oshige, K. Yamaguchi, S.-I. Matsuura, H. Kurita, A. Mizuno, S. Katsura, Anal. Biochem., 400, 145–147 (2010).

[39] H. Nakao, H. Hayashi, T. Yoshino, S. Sugiyama, K. Otobe, T. Ohtani, Nano Lett., 2, 475–479 (2002).

[40] G. Liu, J. Zhao, Langmuir , 22, 2923–2926 (2006).

[41] J. Zhang, Y. Ma, S. Stachura, H. He, Langmuir , 21, 4180–4184 (2005).

[42] Z. Esmail Nazari, L. Gurevich, In preparation.

[43] C.-H. Hsu, C. Chen, M.-L. Jou, A. Y.-L. Lee, Y.-C. Lin, Y.-P. Yu, W.-T. Huang, S.-H. Wu, Nucleic Acids Res., 33, 4053–4064 (2005).

[44] W. Zhang, J. P. Bond, C. F. Anderson, T. M. Lohman, M. T. Record, PNAS, 93, 2511–2516 (1996).

[45] L. Gurevich, T. W. Poulsen, O. Z. Andersen, N. L. Kildeby, P. Fojan, J. Nanosci. Nanotechnol., 10, 1–5 (2010).

[46] P. Fojan, K. Jensen, L. Gurevich, IEEE Xplore 2011, DOI 10.1109/Wireless-vitae.2011.5940906.

[47] D. Nyamjav, A. Ivanisevic, Adv. Mater., 15, 1805–1809 (2003).

[48] D. Nyamjav, A. Ivanisevic, Biomaterials, 26, 2749–2757 (2005).

[49] H. Nakao, M. Gad, S. Sugiyama, K. Otobe, T. Ohtani, J. Am. Chem. Soc., 125, 7162–7163 (2003).

[50] M. Gad, S. Sugiyama, T. Ohtani, J. Biomol. Struct. Dyn., 21, 387–393 (2003).

[51] J. Opitz, F. Braun, R. Seidel, W. Pompe, B. Voit, M. Mertig, Nanotechnology, 15, 717–723 (2004).

[52] H. Kudo, K. Suga, M. Fujihira, Colloids Surf. A., 313, 651–654 (2008).

[53] S. Gad, A. Aurias, N. Puget, A. Mairal, C. Schurra, M. Montagna, S. Pages, V. Caux, S. Mazoyer, A. Bensimon, D. Stoppa-lyonnet, Genes, Chromosomes Cancer, 31, 75–84 (2001).

[54] A. V. de Barros, T. S. Sczepanski, J. Cabrero, J. P. M. Camacho, M. R. Vicari, R. F. Artoni, Aquaculture, 322, 47–50 (2011).

[55] S. Caburet, C. Conti, A. Bensimon, Trends Biotechnol., 20, 344–350 (2002).

[56] D. M. Czajkowsky, J. Liu, J. L. Hamlin, Z. Shao, J. Mol. Biol., 375, 12–19 (2008).

[57] J. N. Bianco, J. Poli, J. Saksouk, J. Bacal, M. J. Silva, K. Yoshida, Y.-L. Lin, H. Tourrière, A. Lengronne, P. Pasero, Methods, 57, 149–157 (2012).

[58] K. Koutroumpas, J. Lygeros, Automatica, 47, 1156–1164 (2011).

[59] R. Lebofsky, R. Heilig, M. Sonnleitner, J. Weissenbach, A. Bensimon, Mol. Biol. Cell., 17, 5337–5345 (2006).

[60] L. Fu, L. Cao, Y. Liu, D. Zhu, Adv. Colloid Interface Sci., 111, 133–157 (2004).

[61] T. H. Labean, H. Li, Nano Today, 2, 26–35 (2007).

[62] H. Li, J. D. Carter, T. H. LaBean, Materials Today, 12, 24–32 (2009).

[63] A. J. Storm, J. van Noort, S. de Vries, C. Dekker, Appl. Phys. Lett., 79, 3881–3883 (2001).

[64] C. Gómez-Navarro, F. Moreno-Herrero, P. J. de Pablo, J. Colchero, J. Gómez-Herrero, A. M. Baró, PNAS, 99, 8484–8487 (2002).

[65] P. J. de Pablo, F. Moreno-Herrero, J. Colchero, J. Gómez Herrero, P. Herrero, A. M. Baró, P. Ordejón, J. M. Soler, E. Artacho, Phys. Rev. Lett., 85, 4992–4995 (2000).

[66] K. Iguchi, Int. J. Mod. Phys. B, 17, 2565–2578 (2003).

[67] H.-W. Fink, C. Schönenberger, Nature, 398, 407–410 (1999).

[68] O. Legrand, D. Côte, U. Bockelmann, Phys. Rev. E, 73, 0319251–6 (2006).

[69] L. Cai, H. Tabata, T. Kawai, Appl. Phys. Lett., 77, 3105–3106 (2000).

[70] Y. Okahata, T. Kobayashi, H. Nakayama, K. Tanaka, Supramol. Sci., 5, 317–320 (1998).

[71] A. Y. Kasumov, M. Kociak, S. Guéron, B. Reulet, V. T. Volkov, D. V Klinov, H. Bouchiat, Science, 291, 280–282 (2001).

[72] M. Taniguchi, T. Kawai, Physica E, 33, 1–12 (2006).

[73] F. Moreno-Herrero, P. Herrero, J. Colchero, C. Gómez-Navarro, J. Gómez-Herrero, A. M. Baró, Nanotechnology, 14, 128–133 (2003).

[74] H. Yang, K. L. Metera, H. F. Sleiman, Coord. Chem. Rev., 254, 2403–2415 (2010).

[75] A. D. Chepelianskii, D. Klinov, A. Kasumov, S. Guéron, O. Pietrement, S. Lyonnais, H. Bouchiat, New J. Phys., 13, 063046 (2011).

[76] M. Bockrath, N. Markovic, A. Shepard, M. Tinkham, L. Gurevich, L. P. Kouwenhoven, M. W. Wu, L. L. Sohn, Nano Lett., 2, 187–190 (2002).

[77] L. Cai, H. Tabata, T. Kawai, Nanotechnology, 12, 211–216 (2001).

[78] P. Maragakis, R. L. Barnett, E. Kaxiras, M. Elstner, T. Frauenheim, Phys. Rev. B, 66, 241104–1–4 (2002).

[79] C.-H. Hsu, C. Chen, M.-L. Jou, A. Y.-L. Lee, Y.-C. Lin, Y.-P. Yu, W.-T. Huang, S.-H. Wu, Nucleic Acids Res., 33, 4053–4064 (2005).

[80] H. Yokota, D. A. Nickerson, B. J. Trask, G. van den Engh, M. Hirst, I. Sadowski, R. Aebersold, Anal. Biochem., 264, 158–164 (1998).

[81] H. Nakao, H. Shiigi, Y. Yamamoto, S. Tokonami, T. Nagaoka, S. Sugiyama, T. Ohtani, Nano Lett., 3, 1391–1394 (2003).

[82] K. Keren, M. Krueger, R. Gilad, G. Ben-Yoseph, U. Sivan, E. Braun, Science, 297, 72–75 (2002).
[83] K. Keren, R. S. Berman, E. Buchstab, U. Sivan, E. Braun, Science, 302, 1380–1382 (2003).
[84] H. Cohen, T. Sapir, N. Borovok, T. Molotsky, R. Di Felice, A. B. Kotlyar, D. Porath, Nano Letters 7, 981–986 (2007).

Biographies

Zeinab E. Nazari received her Master's degree in Pharmacy from Mashhad University of Medical Sciences, Iran in 2007. In 2012, she received her second Master's degree in Nanobiotechnology from Institute of Physics and Nanotechnology, Aalborg University, Denmark. Presently, Zeinab works as research assistant in Leonid Gurevich group. Her research is focused on conductivity measurements of DNA-based nanomaterials.

Leonid Gurevich received his Ph.D. in Physics at the Institute of Solid State Physics (Chernogolovka, Russia) in 1994. He initially worked on high-Tc superconductors but during his postdoc stay at Delft University of Technology became excited about nanotechnology and the possibility of charge transport through a single molecule. Since 2005 he is an Associate Professor at Aalborg University. His research interests focus on molecular electronics, biosensors and nanofabrication.

Online Manuscript Submission

The link for submission is: www.riverpublishers.com/journal

Authors and reviewers can easily set up an account and log in to submit or review papers.

Submission formats for manuscripts: LaTeX, Word, WordPerfect, RTF, TXT.
Submission formats for figures: EPS, TIFF, GIF, JPEG, PPT and Postscript.

LaTeX

For submission in LaTeX, River Publishers has developed a River stylefile, which can be downloaded from http://riverpublishers.com/river_publishers/authors.php

Guidelines for Manuscripts

Please use the Authors' Guidelines for the preparation of manuscripts, which can be downloaded from http://riverpublishers.com/river_publishers/authors.php

In case of difficulties while submitting or other inquiries, please get in touch with us by clicking CONTACT on the journal's site or sending an e-mail to: info@riverpublishers.com

www.ingramcontent.com/pod-product-compliance
Lightning Source LLC
Chambersburg PA
CBHW061323220326

41599CB00026B/5005